SpringerBriefs in Microbiology

For further volumes:
http://www.springer.com/series/8911

Essam Kotb

Fibrinolytic Bacterial Enzymes with Thrombolytic Activity

 Springer

Essam Kotb
Department of Botany and Microbiology
Faculty of Science
Zagazig University
Street 10, 44519 Zagazig
Egypt
e-mail: ekotb@zu.edu.eg

ISSN 2191-5385   e-ISSN 2191-5393
ISBN 978-3-642-24979-2   e-ISBN 978-3-642-24980-8
DOI 10.1007/978-3-642-24980-8
Springer Heidelberg Dordrecht London New York

Library of Congress Control Number: 2011941671

Printed on acid-free paper

Springer is part of Springer Science+Business Media (www.springer.com)

This book is dedicated with profound gratitude to the world's mother, Egypt

Preface

Stress, high blood pressure, smoking, pollution, fast foods, overweight, excessive travelling, surgery, and less movement are common features in our modern life. These features are risky for blood clotting disorders. According to WHO, over 29% of the total mortalities worldwide are due to thrombosis. By the year 2020 cardiovascular diseases (CVDs) may cause an estimated 25 million deaths per year, thus antithrombotic therapy is of great interest.

The available thrombolytic agents such as urokinase are highly expensive, unavailable for commoners, highly antigenic, unspecific, liable, pyretogenic, and hemorrhagenic. Therefore, the production of fibrinolysing enzymes, which was in effect the rapid dissolution of thrombi within the vascular tree, without the last named detriments by microorganisms is considered one of the most influential subjects and is my aim in this book.

The first part of this book throws light on the mechanism of thrombus formation and all thrombolytic medications regimens, focusing on the fibrinolytic enzymes as a new and future promising therapy. The book also throws light on the microbial and non microbial sources of these enzymes, with great scope on the production of these enzymes in vitro from microorganisms, different protocols of enzyme purification and the physical and biochemical characters of these enzymes.

In fact the use of fibrinolytic enzymes in the treatment of thrombi passed with two important stations. The first was with streptokinase and staphylokinase that dissolve thrombi indirectly by activation of endogenous plasmin to degrade fibrin. The second was started in 1987 with the discovery of nattokinase which is able to degrade fibrin directly and efficiently without the previous drawbacks. I think the future research will make great efforts to find other plasmin-like enzymes like nattokinase and increase the enzyme efficacy and fibrin specificity, focusing on developing effective targeted thrombolytic agents by incorporation with thrombus-specific polypeptide or monoclonal antibody.

It is my intent that researchers, pathologists, clinical laboratory scientists, and other physicians serving as laboratory directors will find this book helpful to understand and carry out their responsibilities. Another intent is that residents and fellows will find this book to be a useful tool for learning the basics of assessing of clot lysis abilities of fibrinolytic enzymes and for studying the hemostasis screening tests. I would like to acknowledge my wife, Dr Asmaa, and my children. Their patience and understanding gave me the time and inspiration to research and write this book.

Essam Kotb

Glossary

3,4-DCI	3,4-dichloroisocoumarin
ACE	Angiotensin converting enzyme
AMMP	Armillaria mellea metalloprotease
AOT	Bis(2-ethylhexyl)sodium sulfosuccinate
APSAC	Acylated plasminogen–streptokinase activator complex
APTT	Activated partial thromboplastin time
BAEE	Na-benzoyel-L-arginine ethyl ester
BHI	Brain–heart infusion
BKII	Bacillokinase II
BT	Bleeding Time
CBC	Complete blood count
CCD	Central composite design
CDM	Chemically defined media
CLNE	Na-CBZ-lysine-p-nitrophenyl ester
CU	Plasmin unit
CVDs	Cardiovascular diseases
D	D domain of fibrinogen
DBCLT	Dilute blood clot lysis time
DEAE	Diethylaminoethyl
DFP	Diisopropylfluorophosphate
DIC	Disseminated intravascular coagulation
DIP	Di-isopropyl flurophosphates
DJ	Doen-Jang, a traditional Korean fermented food
DSC	Differential scanning calorimetry
E	E domain of fibrinogen
ECLT	Euglobulin clot lysis time
EDTA	Ethylenediaminetetraacetic acid
EFA	Euglobulin fibrinolytic activity
ELISA	Enzyme-linked immunosorbent assay
EMS	Ethyl methane sulfonate
FDPs	Fibrin degradation products

FFD	Fractional factorial design
Fibrin	Insoluble fibrin
Fibrins	Soluble fibrin
FT-IR	Fourier transform infrared
FU	Fibrin unit
HIT	Heparin-induced thrombocytopen
HMWK	High-molecular-weight kininogen
HPLC	High performance liquid chromatography
IgG	Immunoglobulin type G
INR	International normalizing ratio
IP	Intraperitoneal injection
IPTG	Isopropyl-β-D-thiogalactopyranoside
ITP	Idiopathic thrombocytopenic purpura
IV	Intravenous injection
LB	Luria-Bertani medium
NMR	Nuclear magnetic resonance
NPGB	p-nitrophenyl-p-guanidinobenzoate
PAI-1	Plasminogen activator inhibitor-1
Pas	Plasminogen activators
PEG	Polyethylene glycol
PK	Prekallikrein
PMSF	Phenylmethylsulfonyl fluoride
PSLT	Plasma streptokinase lysis test
PT	Prothrombin time
PTT	Partial-thromboplastin time
Rpm	Rotation per minute
SAK	Staphylokinase
SALT	Streptokinase activated lysis time
SBP	Systolic blood pressure
SGE	Group E streptococci
SK	Streptokinase
SLE	Systemic lupus erythematosus
SMCE	Soybean milk coagulating enzyme
SSF	Solid-state fermentation
SVs	Snake venoms
TAME	Na-tosyl-L-arginine methyl ester
TF	Tissue factor
TLCK	Tosyl-L-lysine chloromethyl ketone
t-PA	Tissue-plasminogen activator
TT	Thrombin time
TTP	Thrombotic thrombocytopenic purpura
U	International unit
UK	Urokinase
VTE	Venous thromboembolism
vWD	Von Willebrand's disease

Contents

Fibrinolytic Bacterial Enzymes with Thrombolytic Activity

Abstract This book describes the fibrinolytic enzymes of microbial origin that are able to dissolve endogenous thrombi in vivo. The fibrinolytic enzyme streptokinase for example, is produced by β-hemolytic streptococci and exerts its enzyme action indirectly by activating plasminogen. On the other hand, staphylokinase is produced by *Staphylococcus aureus* by stoichiometric complexation with plasmin(ogen) that activates other plasminogen molecules. Serrapeptase is a different fibrinolytic enzyme produced by enterobacterium *Serratia* sp. E-15 with multiple functions including fibrin degradation. In addition, nattokinase is a very promising enzyme produced by *Bacillus natto* in fermented soybean in the Japanese diet providing them with the lowest rate of thrombosis disorders all over the world. For each fibrinolytic enzyme there is a special focus on the enzyme structure and mechanism of action. In Sect. 6 the methods used for assessment of clot lysis in vitro are discussed: Fibrin plate methods, streptokinase lysis methods, nephelometric methods, dilute blood clot lysis time, euglobulin lysis time, esterolytic, and fluorimetric assays. Finally, hemostasis screening tests are discussed, such as CBC, PT, PTT, TT, fibrinogen, D-dimer, and BT assays. They should be done regularly to check the physiological and fibrinolytic activity of blood to reduce the onset of endogenous thrombi.

Keywords Fibrinolytic enzymes · Microorganisms · Thrombosis · Hemostasis · Blood clots · Streptokinase · Staphylokinase · Serrapeptase · Nattokinase · Assays

1 Introduction

Enzyme therapies are becoming more prevalent in medicine today, with many manufacturers targeting their advantages in disease treatment. In the last 100 years, enzymes have been increasingly used to treat various diseases.

E. Kotb, *Fibrinolytic Bacterial Enzymes with Thrombolytic Activity*,
SpringerBriefs in Microbiology, DOI: 10.1007/978-3-642-24980-8_1,
© Essam Kotb 2012

Early observations of *Bacillus pyocyaneus* revealed that its secretions could destroy anthrax bacilli and protect mice from inoculation with this deadly bacterium. Scientists deduced that the secretions were able to destroy anthrax via enzymatic degradation. This early observation paved the way for the use of enzymes in medicine. Today, enzymes are used as anticoagulants, oncolytics, thrombolytics, anti-inflammatories, fibrinolytics, mucolytics, antimicrobials, and digestive aids.

Enzymes are found throughout the natural world; the number of uses for them in various fields of industry in addition to medicine is staggering. Enzymes are found in animal and plant sources. Enzymes can be thought of as protein molecules with a specific mission—to initiate and regulate countless biologic reactions in living organisms.

Enzymes are used for metabolic and digestive processes. Metabolic enzymes greatly increase the speed at which chemical processes take place within the body; without enzymes, cells could not perform their multiple functions. Every aspect of life depends on the energetic stimulus that enzymes provide.

Perhaps therapeutic enzymes are used most often for enhancing digestive function. Enzymes help break down food into its smallest components. Enzymes secreted by humans include pepsin and protease for breakdown of proteins, lipase for fats, and amylase for carbohydrates. Cellulase, which helps with digestion of plant cells, is not produced by humans but is extracted from plant tissues as they are mechanically broken down. Plant-based foods are often cooked, but heat destroys enzymes; a plant food in its raw, fresh state produces considerably more enzyme activity than one that has been cooked.

Enzymes, like their application in medicine, exert their effects in a multitude of ways. One primary focus of enzymatic action is on the protein fibrin. Fibrin is an insoluble protein involved in blood clotting. In the many steps of the clotting cascade, fibrin is the final product. It is derived from its soluble protein precursor, fibrinogen. Fibrin is laid down inside blood vessels that have been compromised by disease or injury. Fibrin forms minuscule strands that eventually dry and harden, capturing blood vessel components effectively.

Certainly, fibrin occupies a vital role in health and healing; however, fibrin may also be responsible for an overzealous propensity to form inappropriate clots in the body. Inappropriate clotting, of course, is a major risk factor for myocardial infarction and stroke.

When correctly balanced, deposition and removal of fibrin maintains avoidance of blood loss and adverse viscosity in the vascular system. A balance tipped in favor of fibrin overproduction leads to dangerous clotting.

Various types of thrombosis are responsible for an increasing number of deaths each year. In the USA alone, lung blood clots affect an estimated 1,000,000 patients annually (Lopez-Sendon et al. 1995). According to a report published by the World Health Organization (WHO) in 2001, 17 million people die every year of cardiovascular diseases (CVDs).

The formation of a blood clot in a blood vessel (intravascular thrombosis) is one of the main causes of CVDs. The major protein component of blood clots, fibrin, is formed from fibrinogen via proteolysis by thrombin. Meanwhile, fibrin clots can be hydrolyzed by plasmin to avoid thrombosis in blood vessels. In an unbalanced situation due to some disorders, the clots are not hydrolyzed, and thus thrombosis occurs (Lopez-Sendon et al. 1995).

So, several investigations are being pursued to enhance the efficacy and specificity of fibrinolytic therapy, and microbial fibrinolytic enzymes have attracted much more medical interest in recent decades (Goldhaber and Bounameaux 2001; Tough 2005).

Based on their different working mechanisms, thrombolytic agents are classified into two types. One is plasminogen activators, such as tissue-type plasminogen activator (t-PA) (Collen and Lijnen 2004) and urokinase (Duffy 2002), which activate plasminogen into active plasmin to degrade fibrin. The other type is plasmin-like proteins, which directly degrade fibrin, thereby dissolving thrombi rapidly and completely.

Although plasminogen activators and urokinase are still widely used in thrombolytic therapy today, their expensive prices and undesirable side-effects, such as the risk for internal hemorrhage within the intestinal tract when orally administrated, have prompted researchers to search for cheaper and safer resources (Ismail 1981; Nakajima et al. 1993; Bode et al. 1996). Therefore, microbial fibrinolytic enzymes have also attracted much more medical interest during recent decades (Bode et al. 1996).

Fibrinolytic enzymes were successively discovered from different microorganisms, the most important among which is the genus *Bacillus* from traditional fermented foods (Mine et al. 2005).

The physiochemical properties of these enzymes have been characterized, and their effectiveness in thrombolysis in vivo has been further identified. Therefore, microbial fibrinolytic enzymes, especially those from food-grade microorganisms, have potential to be developed as functional food additives and drugs to prevent or cure thrombosis and other related diseases (Ambrus et al. 1979; Sumi et al. 1987, 1990; Kim et al. 1996a, b; Hwang et al. 2002).

Fibrinolytic enzymes are mainly proteases. These catalyze total hydrolysis of proteins and specifically act on interior peptide bonds (Bayoudh et al. 2000).

All living cells produce different types of proteases, but the majority are produced by microorganisms. Many workers have reported that bacteria are high protease producers (Kalisz 1988).

Proteases are grossly subdivided into two major groups, namely exopeptidases and endopeptidases, depending on their site of action. Exopeptidases cleave the peptide bond proximal to the amino or carboxy termini of the substrate, whereas endopeptidases cleave peptide bonds distant from the termini of the substrate (International Union of Biochemistry and Molecular Biology 1992).

Based on the functional group present at the active site, proteases are further classified into four families: serine proteases, aspartic proteases, cysteine proteases, and metalloproteases (Hartley 1960).

Fibrinolytic proteases are mainly serine or metalloproteases (Sharma et al. 2004) and are also of major importance in food, leather, detergent, pharmaceutical, and

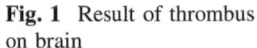

Fig. 1 Result of thrombus on brain

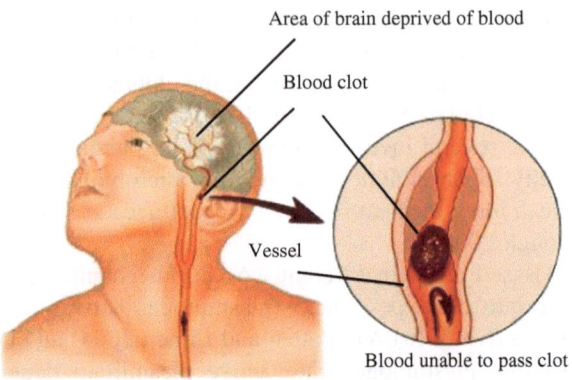

Area of brain deprived of blood

Blood clot

Vessel

Blood unable to pass clot

waste management industries. Proteases constitute two-thirds of the total number of enzymes used in industry, and this is expected to increase (Gupta et al. 2002).

1.1 Thrombus Formation

Blood clots form when fibrin strands accumulate in a blood vessel. In the heart, blood clots cause blockage of blood flow to muscle tissue. If blood flow is blocked, oxygen supply to that tissue is cut off, and it eventually dies (Fig. 1). This can result in angina and heart attacks. Thrombi in chambers of the heart can mobilize to the brain. In the brain, blood clots also block blood and oxygen from reaching necessary areas, which can result in senility and/or stroke.

Thrombolytic enzymes are normally generated in endothelial cells of blood vessels. As the body ages, production of these enzymes begins to decline, making blood more prone to coagulation. This mechanism can lead to cardiac or cerebral infarction, as well as other conditions. Since endothelial cells exist throughout the body, such as in the arteries, veins, and lymphatic system, poor production of thrombolytic enzymes can lead to the development of thrombotic conditions virtually anywhere in the body.

It has recently been revealed that thrombotic clogging of cerebral blood vessels may be a cause of dementia. It has been estimated that senile dementia is caused by thrombus in 60% of such patients in Japan. Thrombotic diseases typically include cerebral hemorrhage, cerebral infarction, cardiac infarction, and angina pectoris, and also include diseases caused by blood vessels with lowered flexibility, including senile dementia and diabetes (caused by pancreatic dysfunction). Hemorrhoids are considered a local thrombotic condition. If chronic diseases of the capillaries are also considered, the number of thrombus-related conditions may be much higher. Cardiac infarction patients may have an inherent imbalance in that their thrombolytic enzymes are weaker than their coagulant enzymes. Nattokinase holds great promise to support patients with such inherent weaknesses in a convenient and consistent manner, without side-effects (Health Canada 2000).

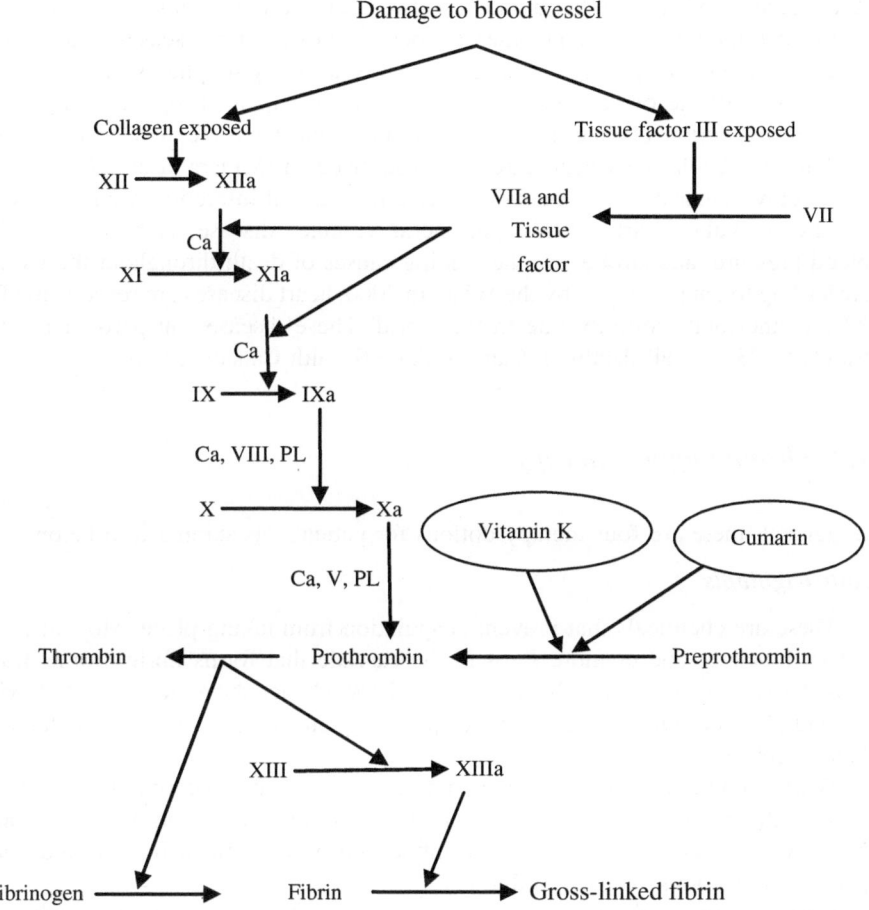

Fig. 2 The blood-clotting cascade

Nattokinase is capable of directly and potently decomposing fibrin as well as activating pro-urokinase (endogenous).

In fact, hemostasis is the tightly regulated process of keeping an optimal balance between coagulation and anticoagulation. Coagulation involves a series of enzymatic reactions, in which inactive plasma proteins are converted into active enzymes in each step of the pathway. As shown in Fig. 2, the cascade is initiated by the release of tissue factor or damaged collagen underneath the blood vessel endothelium. The final step involves the formation of a fibrin clot that stabilizes the platelet plug. The fibrin clot is formed from fibrinogen by thrombin. The dissolution of a blood clot is dependent on the action of endogenous plasmin, a serine protease that is activated by tissue plasminogen activator (Silverthorn et al. 1998).

An imbalance in hemostasis may result in excessive bleeding, or formation of a thrombus (blood clot) that adheres to the unbroken wall of the blood vessels.

Accumulation of fibrin in the blood vessels can interfere with blood flow and lead to myocardial infarction and other serious cardiovascular diseases. Unless the blockage is removed promptly, the tissue that is normally supplied with oxygen by the vessel will die or be severely damaged. If the damaged region is large, the normal conduction of electrical signals through the ventricle will be disrupted, leading to irregular heartbeat, cardiac arrest, or death (Mihara et al. 1991).

Cardiovascular diseases, including acute myocardial infarction, ischemic heart disease, valvular heart disease, peripheral vascular disease, arrhythmias, high blood pressure, and stroke, are the leading causes of death throughout the world. According to data provided by the WHO in 2000, heart diseases are responsible for 29% of the total mortality rate in the world. These diseases, in particular, contribute to 38% of all deaths in Canada alone (Health Canada 2000).

1.2 Thrombolytic Therapy

In general, there are four therapy options for patients, as summarized below.

Anticoagulants

These are chemicals that prevent coagulation from taking place. Most of them act by blocking one or more steps in the cascade that forms fibrin. Some drugs inhibit synthesis of clotting factors, while others enhance the anticoagulant activity of naturally occurring blood factors or prevent platelet plug formation (Oden and Fahlen 2002).

Warfarin inhibits coagulation by interfering with the incorporation of vitamin K into vitamin K-dependent clotting factors, including factors II, VII, IX, and X. However, its effectiveness can be influenced by age, racial background, diet, and co-medications such as antibiotics.

Heparin is another example of an anticoagulant, whose major effect is inhibition of thrombin, factor IIa, and factor Xa in the coagulation cascade. It has a short half-life, being associated with hemorrhage, osteoporosis, alopecia, thrombocytopenia, and hypersensitivity (Fitzmaurice et al. 2002).

Antiplatelets

These are used to prevent a clot from forming or from getting larger and occluding the entire vessel.

Aspirin is the most widely used anti-platelet drug, and it inhibits platelet aggregation. Adverse effects of aspirin are similar to those of warfarin.

Other anti-platelet drugs, such as dipyridamole, clopidogrel, and ticlopidine, work by inhibiting platelet-activating factor and collagen. Their usage, however, is associated with bone marrow suppression, in particular leukopenia (Blann et al. 2002).

Mechanical and Surgical Treatments

These are usually reserved for massive pulmonary embolism where drug treatments have failed (Wheatley 2002). In fact, surgeons have developed new strategies to maximize the effectiveness of coronary surgery. However, these require use of an anticoagulant as adjuvant therapy (Turpie et al. 2002).

Fibrinolytic Enzymes

Unlike heparin and warfarin, fibrinolytic enzymes lyse pre-existing thrombi. These include urokinase (extracted from kidneys), streptokinase (extracted from bacteria), and genetically engineered tissue plasminogen activator (t-PA). Evidence has shown that patients with pulmonary embolism treated with streptokinase and urokinase are three times more likely to show clot resolution than patients taking heparin alone. These enzymes can also prevent some damage if the clot is removed soon after it occurs.

Streptokinase is an effective thrombolytic agent derived from streptococci. It can potentiate the body's own fibrinolytic pathways by converting plasminogen to plasmin. Being bacteria-derived, it is antigenic and may result in the development of neutralizing antibodies and allergic reactions.

On the other hand, t-PA is produced by recombinant DNA technology; it mimics an endogenous molecule that activates the fibrinolytic system. It does not elicit an allergic response and is considered to be more clot specific (Mine et al. 2005).

The above fibrinolytic enzymes lack site specificity and have adverse effects such as gastrointestinal hemorrhage (Turpie et al. 2002); systemic fibrinogenolysis with accompanying bleeding is frequently encountered, as well as neurological complications (Caramelli et al. 1992), including stroke (Maggioni et al. 1992) and intracranial hemorrhage (Kase et al. 1992).

The limited efficacy and potentially life-threatening side-effects of available thrombolytic agents remain problems. Investigators have attempted to overcome these problems by enhancing thrombolytic activity and improving targeting to the clot. Despite these modifications, all current thrombolytic agents depend on plasmin being generated through plasminogen activation (indirect action). Thus, there is a strong rationale for evaluating newer agents that act directly on fibrin to convert it to fibrin degradation products (FDP) (Caramelli et al. 1992). This is achieved by fibrinolytic agents from living organisms.

1.3 Sources of Fibrinolytic Enzymes

From Bacteria

Bacillus subtilis strain natto produces a highly potent fibrinolytic enzyme called nattokinase that was first screened from a traditional Japanese soybean-fermented food (Tables 1, 2) named *natto* (Sumi et al. 1987).

Table 1 Bacilli from traditional food

Microorganism	Food	Name of enzyme	References
B. natto, NK	*Natto*, Japan	Nattokinase	Fujita et al. (1993)
B. amyloliquefaciens DC-4	*Douchi*, China	Subtilisin DFE	Peng et al. (2003)
Bacillus sp. CK	*Chungkook-jang*, Korea	CK	Kim et al. (1996a, b)
Bacillus sp. DJ-4	*Doen-jang*, Korea	Subtilisin DJ-4	Kim and Choi (2000)
Bacillus sp. DJ-2	*Doen-jang*, Korea	bpDJ-2	Choi et al. (2005)
Bacillus sp. KA38	*Jeot-gal*, Korea	*Jeot-gal* enzyme	Kim et al. (1997)
B. subtilis QK02	Fermented soybean	QK-1 and QK-2	Ko et al. (2004)
Bacillus firmus NA-1	*Natto*	–	Seo and Lee (2004)
B. subtilis IMR-NK1	*Natto*	–	Chang et al. (2000)
Bacillus sp.	*Tofuyo*, Japan	SMCE	Fujita et al. (1993)
Katsuwonus pelamis	Skipjack, Japan	Katsuwokinase	Sumi et al. (1995)
Bacillus sp.	*Kimchi*, Korea	Bacillus protease	Noh et al. (1999)
Armillaria mella	*Armillaria mella*	Neutral metalloprotease	Kim and Kim (1999)
Bacillus sp. KDO-13	Soybean paste, Korea	–	Lee et al. (2001)

Table 2 Sources of microbial fibrinolytic enzymes

Microorganism	References
Bacilli	
B. subtilis BK-17	Jeong et al. (2001)
B. subtilis A1	Jeong et al. (2004)
B. subtilis 168	Kho et al. (2005)
Actinomyces thermovulgaris	Egorov et al. (1976)
Streptomyces	
S. megasporus SD5	Chitte and Dey (2000, 2002)
S. spheroids M8-2	Egorov et al. (1985)
Streptomyces sp. Y405	Wang et al. (1999a)
Fungi	
A. ochraceus 513	Batomunkueva and Egorov (2001)
Cochliobolus lunatus	Abdel-Fattah and Ismail (1984)
F. oxysporum	Tao et al. (1997, 1998)
Fusarium pallidoroseum	El-Aassar (1995)
P. chrysogenum H9	El-Aassar et al. (1990)
Pleurotus ostreatus	Choi and Shin (1998)
R. chinensis 12	Xiao-Lan et al. (2005)
Algae	
C. intricatum	Matsubara et al. (1998)
C. latum	Matsubara et al. (1999)
C. divaricatum	Matsubara et al. (2000)

Streptococcus hemolyticus and *Staph aureus* produce streptokinase and staphylokinase that were earlier proved to be effective in thrombolytic therapy (Collen and Lijnen 1994).

Some other bacilli from different fermented foods (Table 1) were discovered to produce fibrinolytic enzymes (Table 2). *Bacillus* sp. CK from the Korean fermented soybean sauce named *chungkook-jang* (Kim et al. 1996a, b), *Bacillus* sp. KA38 from the Korean salty fermented fish called *jeot-gal* (Kim et al. 1997), and *B. amyloliquefaciens* DC-4 from the Chinese soybean-fermented food named *douchi* (Peng and Zhang 2002a).

Subtilisin DFE was then screened from *Bacillus* sp. strains DJ-2 (Kim and Choi 2000) and subtilisin DJ-4 from *Bacillus* sp. strains DJ-4 from the Korean *doen-jang* (Choi et al. 2005).

Yoon et al. (2002) systematically screened the fibrinolytic enzyme-producing strains from many commercial and home made fermented foods including *natto*, *chungkook-jang*, *doen-jang*, *jeot-gal*, and the Indonesian fermented food *tempeh*, and successfully isolated the strain *Enterococcus faecalis* producing higher fibrinolytic activity.

These exciting findings imply the possibility of consuming fermented foods to prevent cardiovascular diseases.

Suzuki et al. (2003a) reported that dietary supplementation with *natto* could shorten euglobulin clot lysis time, which is used to evaluate the total intrinsic fibrinolytic activity in plasma. At the same time, dietary *natto* extract did not prolong bleeding time, indicating the safety of *natto* to be developed as a functional food.

The vegetable cheese-like food *natto* is extremely popular in Japan, with a history extending back over 1,000 years. A fibrinolytic enzyme, termed nattokinase, can be extracted from *natto*; the enzyme is a subtilisin-like serine protease composed of 275 amino acid residues and has molecular weight of 27.7 kDa. In vitro and in vivo studies have consistently demonstrated the potent profibrinolytic effect of the enzyme (Pais et al. 2006).

In addition, Wang et al. (2009) showed that *Bacillus* spp. produced a variety of extracellular and intracellular fibrinolytic enzymes (nattokinases). They also reported that nattokinase could be purified from culture supernatant of *Pseudomonas* sp. TKU015 isolated from soils, exhibiting high activity towards fibrin.

From Other Microorganisms

Streptomyces megasporus SD5, isolated from the water of a hot spring, could produce a strong thermostable fibrinolytic enzyme (Chitte and Dey 2000).

Some kinds of fungi have also been found to produce proteases with high fibrinolytic activity, for example, *Penicillium chrysogenum* (El-Aassar et al. 1990), *Fusarium oxysporum* (Sun et al. 1998), *Aspergillus ochraceus* 513 (Batomunkueva and Egorov 2001), and *Rhizopus chinensis* 12 (Xiao-Lan et al. 2005).

Many fibrinolytic enzymes have also been identified in the fruiting bodies and culture supernatants of different medicinal mushrooms, such as *Grifora frondosa* aminopeptidase (Nonaka et al. 1997), *Pleurotus ostreatus* metalloprotease (Choi and Shin 1998), *Armillaria mellea* metalloprotease (Kim and Kim 1998; Healy et al. 1999), *Perenniporia fraxinea* proteases 1 and 2 (Lee et al. 2006), and *Cordyceps militaris* (Cui et al. 2008).

Matsubara et al. (1998, 1999, 2000) discovered fibrinolytic enzymes from marine algae *Codium latum*, *Codium divaricatum*, and *Codium intricatum* (Table 2).

From Other Organisms

Fibrinolytic enzymes have been found in the hemorrhagic toxin of snake venoms (Nikai et al. 1984). Snake venoms (SVs) contain many biologically active components that affect hemostasis (White 2005). Fibrinolytic enzymes have been isolated from venoms of Viperidae, Elapidae, and Crotalidae snakes (Markland 1998; Swenson and Markland 2005). The majority of SV fibrinolytic enzymes are zinc metalloproteinases.

Proteinase activities in earthworms have long been studied. *Eisenia fetida* showed strong fibrinolytic and hemolytic activity (Roch 1979; Milochau et al. 1997; Yang and Ru 1997; Hrženjak et al. 1998; Tang et al. 2002; Wang et al. 2003). In addition, Mihara et al. (1983) have demonstrated a novel fibrinolytic enzyme extracted from *Lumbricus rubellus*, named lumbrokinase.

Sumi et al. (1993) and Nakajima et al. (1993, 1996) purified and characterized some potent fibrinolytic enzymes in *L. rubellus*. In addition, Xu et al. (2002) purified a fibrinolytic enzyme in *L. bimastus*.

Fibrinolytic enzymes with thrombolytic activity have been also isolated from vampire bat (Cartwright 1974; Gardell et al. 1989), egg cases of praying mantis *Tenodera sinensis* (Wang et al. 1989), dung beetles, *Catharsius molossus* (Ahn et al. 2003), and the herbal medicines *Spirodela polyrhiza* (Choi and Sa 2001) and the seaweed *Codiales codium* (Jeon et al. 1995).

Construction of Genetically Engineered Strains

Conventional mutagenesis and current recombinant DNA technology have been adopted to improve enzyme production and simplify downstream manipulation. For the former, random mutagenesis is combined with high-throughout screening methods to isolate the expected mutant strains. Recently, Lai et al. (2004) reported successful doubling of the specific activity of fibrinolytic enzyme through random mutagenesis in vitro using the chemical ethyl methane sulfonate (EMS).

In molecular biology applications, *Bacillus subtilis* has been recognized as a good host for expression of foreign proteins with pharmacological activities, because of its nonpathogenicity and capability of secreting functional extracellular proteins to the culture medium (Wong 1995). Subtilisin DFE was actively expressed in the protease-deficient strain *B. subtilis* WB600 (Peng et al. 2004). Furthermore, the native promoter of subtilisin DFE gene was replaced by that of the α-amylase gene from *B. amyloliquefaciens* DC-4, resulting in a sharp increase in fibrinolytic activity from 80 to 200 urokinase units per milliliter

(Xiao et al. 2004). Liu and Song (2002) succeeded in the functional expression of NK in *B. subtilis* as well.

When expressed in *E. coli*, NK and subtilisin DFE formed insoluble aggregates without enzymatic activity (Peng and Zhang 2002c). However, two newly published papers reported successful expression of active NK and subtilisin DFE in *E. coli* (Chiang et al. 2005; Zhang et al. 2005). Both studies took advantage of the principle that the extracellular protease subtilisin from the genus *Bacillus* is synthesized as pre-proenzymes, and its propeptide may function as an intramolecular chaperone to facilitate correct folding of the protease domain (Ikemura and Inouye 1988; Zhu et al. 1989). Zhang et al. (2005) showed that subtilisin DFE was highly expressed in *E. coli* BL21 (DE3) as a fusion protein of Trx–prosubtilisin DFE via the expression vector pET32a and that strong fibrinolytic activity was detected in both soluble fraction and inclusion bodies fraction after in vitro renaturation. Moreover, the fusion proteins are easily purified and refolded in a column to active enzyme. Most importantly, Trx–propeptide can be automatically cleaved during in vitro refolding to form mature subtilisin DFE. Chiang et al. (2005) indicated that either nattokinase or pronattokinase could be overexpressed in *E. coli* as a recombinant protein fused to the C-terminus of olesin, a unique structural protein of seed oil bodies, by a linker polypeptide intein. After reconstitution of artificial oil bodies, active NK was released through self-splicing of intein induced by temperature alteration and spontaneous cleavage of the propeptide (Chiang et al. 2005).

The Human Plasma Fibrinolytic System

The human plasma fibrinolytic system is capable of degradation and dissolution of fibrin clots by the proteolytic enzyme plasmin, which is formed by activation of an inactive plasma zymogen precursor, plasminogen, by specific plasma or tissue activators (Robbins 1978).

Between 1965 and 1970, a highly purified zymogen was prepared from enriched human plasma fractions (Robbins and Summaria 1970, 1976) for treatment of blood clots. Plasminogen activators occur naturally in many different tissues, in body fluids, and in blood (Robbins 1978).

1.4 Production of Bacterial Fibrinolytic Enzyme In Vitro

The cost of enzyme production and downstream processing is the major obstacle to successful application of proteases in industry and medicine. For fibrinolytic enzymes, many attempts have been done to improve expression of the fibrinolytic enzyme, including selection of an ideal culture medium, optimization of environmental conditions, and overexpression by genetically engineered strains.

Selection of medium components is usually critical for fermentative production of fibrinolytic enzymes. Since different microorganisms possess diverse physiological characteristics, it is necessary to optimize nutrient components and environmental conditions for cell growth and fibrinolytic enzyme production.

Liquid fermentation is usually considered as the first choice for bacteria, whereas solid-state fermentation (SSF) is favored for fungi. SSF has numerous advantages for enzyme production, such as low wastewater output, low operating costs, and high productivity.

Tao et al. (1997, 1998) systematically studied production of the fibrinolytic enzyme from *F. oxysporum* using different SSF methods, largely increasing production and reducing the cost; for instance, soluble starch or dextrin was the best carbon source for *B. amyloliquefaciens* DC-4 due to the strong amylase activity (Peng and Zhang 2002b).

The optimal temperature of *Streptomyces megasporus* SD5 for enzyme synthesis is 55°C, because the strain was isolated from a hot spring (Chitte and Dey 2002). In some cases, fibrin was found to enhance the enzyme production, suggesting that fibrin, as a substrate of fibrinolytic enzyme, could activate or induce enzyme production during cultivation (Chitte and Dey 2002; Peng and Zhang 2002b).

Liu et al. (2005) employed the statistical methods of fractional factorial design (FFD) and central composite design (CCD) to optimize the fermentation media for production of NK and finally increased the fibrinolytic activity to 1,300 U/ml, about 6.5 times higher than the original value (Fu et al. 1997).

Although the traditional one-at-a-time optimization strategy is simple and easy, it frequently fails to locate the region of optimal response, because the comprehensive effect of factors is not taken into consideration.

Conventional mutagenesis and current recombinant DNA technology have also been adopted to improve enzyme production. Recently, Lai et al. (2004) showed successful doubling of the specific activity of fibrinolytic enzyme through random mutagenesis in vitro using EMS.

In molecular biology applications, *Bacillus subtilis* has been recognized as a good host for expression of foreign proteins with pharmacological activities because of its nonpathogenicity and capability of secreting functional extracellular proteins to the culture medium (Wong 1995).

The optimum temperature for production of fibrinolytic enzymes varies according to the kind of bacterium, the substrate used, as well as other environmental conditions during fermentation. Active production of the enzyme by *B. subtilis* LD.8547 was between 35°C and 65°C, with optimum activity at 50°C (Wang et al. 2008), while the enzyme production by the marine bacterium *B. subtilis* A26 was maximum at temperature from 50°C to 70°C, with optimum of 60°C (Agrebi et al. 2009).

The fermentation period plays an essential part in the process of enzyme production. It varies according to the producing strain and other environmental conditions. Nattokinase activity produced by *Pseudomonas* sp. TKU015 was found at the second day and then decreased gradually (Wang et al. 2009).

The effect of initial pH on the production of proteolytic enzymes by various microorganisms has been studied by several workers. The maximum production of the enzyme by *B. subilis* LD.8547 was at a pH range of 7.0–10.0, with pH 8.0 as the optimum pH, while Ashipala and He (2008) reported pH 6.52 as the best for fibrinolytic enzyme production by *Bacillus subtilis* DC-2, using statistical programs.

Several workers have discussed the importance of various sugars for production of bacterial proteases. Ismail et al. (2004) stated that starch in the basal medium played an active role in biosynthesis of fibrinolytic enzyme by *B. macerans* 3185. Starch concentration of 0.2% (w/v) was sufficient to achieve the maximal productivity. On the other hand, Liu et al. (2005) reported that the optimal carbon source for nattokinase production by *B. natto* NLSSE was maltose. Also, glucose and sucrose had similar positive effects, while xylose and glycerol performed poorly. In addition, Deepak et al. (2008) reported glucose as the best C-source for *Bacillus subtilis*.

Several workers have investigated the effect of nitrogen sources on protease production. Liu et al. (2005) reported that, when inorganic nitrogen sources were used, very poor enzyme activities were achieved, and much higher activities were obtained with organic nitrogen sources, with soya bean being the optimal nitrogen source. They also added that organic nitrogen sources (casein, soypeptone, and soya meal) could bring out glutamine and other amino acids by enzyme catalysis, all of which were preferred nitrogen sources of *B. subtilis*.

1.5 Enzyme Purification

The culture filtrate containing the enzyme also contains hundreds of (or more) other proteins of both large and small molecular weight. To study a given enzyme property clearly, it must be purified from other contaminants (Dixon and Webb 1964).

Recovery of fibrinolytic proteases is achieved by salting out the enzyme from its aqueous solution by means of a suitable electrolyte such as ammonium sulfate or sodium chloride. Also, it can be precipitated from the broth filtrate by addition of organic solvents that are miscible with water but not solvents for the enzyme, such as lower alkanes and alkenes. The precipitates obtained usually are amorphous, gummy masses, not easily separated from the aqueous medium (Bayoudh et al. 2000).

Kim et al. (1997) purified *Bacillus* sp. KA38 fibrinolytic enzyme in three steps: ammonium sulfate treatment, diethylaminoethyl cellulose (DEAE) column, and Mono Q column chromatography. The purity of the enzyme solution was further improved by liquid column chromatography.

Jeong et al. (2001) reported that *Bacillus subtilis* BK-17 fibrinolytic enzyme was best purified by applying 75% ethanol saturation followed by loading on to a DEAE-Sephadex A-50, then subjecting the eluates to Sephadex G-75 column chromatography.

Liu et al. (2004) purified nattokinase from *Bacillus natto* NLSSE by liquid/liquid chromatography using AOT/isooctane micellar solution as an extractant.

On the over hand, Lee et al. (2005) purified *Armillaria mellea* fibrinolytic protease by adding an equal volume of pre-chilled ethanol to the culture filtrate up to 75% alcohol saturation, applying it to a carboxymethyl (CM)-cellulose column, and finally applied gel filtration with Sephadex G-75 column chromatography.

1.6 Biochemical Characteristics of Purified Microbial Fibrinolytic Enzymes

Some microbial fibrinolytic enzymes including those from *Streptomyces* (Wang et al. 1999), *R. chinensis* (Liu et al. 2005), *Armillaria mellea* (Lee et al. 2005), and genus *Bacillus* (Jeong et al. 2001; Paik et al. 2004) have been purified and characterized. Their biochemical properties, such as optimal pH and temperature, stability, molecular weight, and substrate specificity, are summarized in Table 3. Some N-terminal sequences have been determined as well (Table 4). According to their catalytic mechanisms, these enzymes are classified into serine proteases (NK, subtilisin DFE, and CK) and metalloproteases (*jeot-gal* enzyme, AMMP, and bacillokinase II), except for those from *R. chinensis* 12 and *Streptomyces* sp. Y405, which are both serine and metalloprotease (Liu et al. 2005; Wang et al. 1999).

The fibrinolytic enzymes belonging to the serine proteases are characterized by the presence of a serine group in their active site. Based on their structural similarities, serine proteases have been grouped into 20 families. Serine proteases are recognized by their irreversible inhibition by 3,4-dichloroisocoumarin (3,4-DCI), diisopropyl fluorophosphate (DFP), phenylmethylsulfonyl fluoride (PMSF), and tosyl-lysine chloromethyl ketone (TLCK).

They are generally active at neutral and alkaline pH, with an optimum between pH 7 and 11. Their molecular masses range between 18 and 35 kDa. The isoelectric points of serine proteases are generally between pH 4 and 6 (Govind et al. 1981).

Most known fibrinolytic enzymes from microorganisms are serine proteases, such as the different fibrinolytic enzymes from several strains of *Bacillus subtilis* (Chang et al. 2000; Ko et al. 2004; Kim et al. 1996a, b; Wang et al. 2006a, b; Wang et al. 2008; Agrebi et al. 2009), subtilisin DJ-4 secreted by *Bacillus* sp. DJ-4 screened from *doen-jang*, a traditional Korean fermented food (Kim and Choi 2000), the fibrinolytic enzyme (subtilisin DFE) produced by *Bacillus amyloliquefaciens* DC-4 screened from *douchi*, a traditional Chinese soybean food (Peng et al. 2003), the two novel fibrinolytic enzymes from endophytic strain *Paenibacillus polymyxa* with relative molecular weight larger than known fibrinolytic enzymes (Lu et al. 2007), and the nattokinase produced by *Pseudomonas* sp. TKU015 (Wang et al. 2009).

However, the fibrinolytic enzymes belonging to the metalloproteases are the most diverse of the catalytic types of proteases. They are characterized by the requirement for a divalent metal ion for their activity. About 30 families of metalloproteases have been recognized based on the nature of the amino acid that completes the metal-binding site. All of them are inhibited by chelating agents such as ethylenediaminetetraacetic acid (EDTA) but not by sulfhydryl agents or DFP (Barett 1995).

Several fibrinolytic metalloproteases are present among microorganisms. Kim et al. (1997) reported a fibrinolytic metalloprotease from fish fermented by *Bacillus* sp. KA38. Jeong et al. (2004) reported a fibrinolytic metalloprotease from *Bacillus subtilis* strain A1, with optimal activity at 50°C. Chiang et al. (2005) reported that

Table 3 Properties of microbial fibrinolytic enzymes

Enzyme, microorganism	Mol. wt, pI, optimal pH, and temperature	pH, temperature stability	Substrate specificity	Comments	References
Nattokinase, *B. natto*	27.7 kDa, pI 8.6	Stable at pH 7–12 and less than 50°C	Best synthetic substrate is that for plasmin or subtilisin	Subtilisin-family serine protease	Fujita et al. (1993), Sumi et al. (1987)
Subtilisin DFE, *B. amyloliquefaciens* DC-4	28 kDa, pI 8.0, pH 9.0, 48°C	Stable at pH 6–10 and less than 50°C for 60 min	Best synthetic substrate is that for subtilisin. Does not degrade blood cells	Subtilisin-family serine protease	Peng et al. (2003)
CK, *Bacillus* sp. CK	28.2 kDa, pH 10, 70°C	Stable at pH 7–10.5, less than 50°C for 60 min	CK has eightfold higher fibrinolytic activity than that of subtilisin Carlsberg Best synthetic substrate is that for plasmin	Thermophilic alkaline serine protease	Kim et al. (1996a, b)
Subtilisin DJ-4, *Bacillus* sp. DJ-4	29 kDa, pH 10.0, 40°C	Stable at pH 4–11 at room temperature for 48 h	Fibrinolytic activity 2.2 and 4.3 times higher than those of subtilisin BPN and subtilisin Carlsberg	Plasmin-like serine protease	Kim and Choi (2000)
Subtilisin QK-2, *B. subtilis* QK02	28 kDa, pH 8.5, 55°C	Stable at pH 3–12 and at 40°C for 30 min	Stronger activity towards the substrate of subtilisin	Subtilisin-family serine protease	Ko et al. (2004)
BK-17, *B. subtilis* BK-17	31 kDa substrate		Most sensitive synthetic is that for plasmin	Directly degrades fibrin, or forms plasmin from plasminogen to degrade fibrin	Jeong et al. (2001)

(continued)

Table 3 (continued)

Enzyme, microorganism	Mol. wt., pI, optimal pH, and temperature	pH, temperature stability	Substrate specificity	Comments	References
From KCK-7, *B. subtilis* KCK-7	44 kDa, pH 8.0, 50°C	Stable up to 60°C over pH 7–10		Serine protease. Ca^{2+} and Cu^{2+} enhance its activity	Paik et al. (2004)
Jeot-gal enzyme, *Bacillus* sp. KA38	41 kDa, pH 7.0, 40°C	Stable up to 40°C and at pH 7–9	Degrades fibrin more easily than α-casein and skim milk. Does not degrade red cells	Metalloprotease, possibly with Zn^{2+} on its active site	Kim et al. (1997)
Bacillokinase II (BKII), *B. subtilis* A1	31.4 kDa, pH 7.0, 50°C	Stable at pH 4.0 and up to 50°C	Highest amidolytic activity on the synthetic substrate for kallikrein	Metalloprotease. Negligible activity on skim milk or gelatin and no activity on red blood cells	Jeong et al. (2004)
AMMP, *Armillaria mellea*	21 kDa, pH 6.0, 33°C		Preferentially hydrolyzes Aα fibrinogen chain over Bβ and γ chains	Chymotrypsin-like metalloprotease. Ca^{2+} and Mg^{2+} enhance its activity	Lee et al. (2005)
From *Bacillus* sp. KDO-13	45 kDa, pH 7.0, 60°C	Stable at pH 7–9 and 50°C	Low K_m value for fibrin hydrolysis	Metalloprotease, Co^{2+} and Hg^{2+} enhance its activity	Lee et al. (2001)
From *R. chinensis* 12	18 kDa, pI 8.5, pH 10.5, 45°C	Stable at pH 6.8–8.8 at 37°C for 24 h	Degrades fibrin, also cleaves α, β, γ chains of fibrinogen simultaneously	Hydrosulfuryl metalloprotease	Xiao-Lan et al. (2005)
SW-1, *Streptomyces* sp. Y405	30 kDa, pI 8.5, pH 8.0	Stable at 4–37°C, pH 4–9	Degrades fibrin directly	Serine and metalloprotease	Wang et al. (1999a)

Table 4 Comparison of N-terminal amino acid sequence of microbial fibrinolytic enzymes

Enzyme	N-terminal amino acid sequence	Reference
NK	AQSVPYGISQIKAPALHSQGYTGS	Fujita et al. (1993)
Subtilisin QK-2	AQSVPYGISQIKAPALHSQG	Ko et al. (2004)
Subtilisin DFE	AQSVPYGVSQIKAPALHSQGFTGS	Peng et al. (2003)
Subtilisin DJ-4	AQSVPYGVSQIKAP	Kim and Choi (2000)
31-kDa enzyme	AQSVPYGVSQIKAPAAHN	Jeong et al. (2001)
Chungkook-Jang "CK"	AQTVPYGIPLIKAD	Kim et al. (1996a, b)
Tofuyo "SMCE"	AQTVPYGIPQIKAD	Kim and Kim (1999)
Bacillokinase II	ARAGEALRDIYD	Jeong et al. (2004)
KA38	VYPFPGPIPN	Kim et al. (1997)
CLP	VVGGDEPP	Matsubara et al. (1999)
CIP	X-TPLTQVLSGNAVLVEAVLVEAVKA	Matsubara et al. (2000)
Katsuwokinase Skipjack "Shiokara"	IVGGYEQZAHSQPHQ	Sumi et al. (1995)
SW-1	R/N/F–P/D-GMTMTAIANQNTQIN	Wang et al. (1999a, b)
Fermented shrimp paste	DPYEEPGPCENLQVA	Mine et al. (2005)
AMMP	MFSLSSRFFLYTLCLSAVAVSAAP	Lee et al. (2005)
bpDJ-2	TDGVEWNVDQIDAPKAW	Choi et al. (2005)

Vibrio vulnificus secretes a broad-specific metalloprotease capable of interfering with blood hemostasis. Additionally, the fibrinolytic enzyme from *R. chinensis* 12 was found to be a hydrosulfuryl-metalloprotease (Xiao-Lan et al. 2005).

All these fibrinolytic enzymes have high substrate specificity to fibrin, different from other proteases with broad substrate specificity; for instance, CK activity for degrading fibrin is about eight times higher than that of subtilisin Carlsberg, a common alkaline protease with identical N-terminal sequence (Kim et al. 1996a, b). Similar examples also include NK and subtilisin E (Nakamura et al. 1992), subtilisin DFE and subtilisin BPN (Peng et al. 2004), and subtilisin DJ-4 and subtilisin BPN (Choi et al. 2004). Although these common alkaline proteases (subtilisin E, BPN, and Carlsberg) have highly homologous sequence with the corresponding fibrinolytic enzymes, why is it that only these fibrinolytic enzymes have very high substrate specificity to fibrin? The evolutionary changes of the critical amino acid residues in the substrate binding site probably account for this difference. However, more research should be done to completely elucidate this interesting phenomenon and provide some hints of the structure–function relationship. Furthermore, some microbial fibrinolytic enzymes can activate plasminogen to further enhance fibrinolysis (Kim et al. 1996a, b).

2 Streptokinase

Streptokinase is the best known microbial plasminogen activator. Staphylokinase (SAK) sourced from *Staphylococcus* sp. is a potential alternative plasminogen activator (Okada et al. 2001). Recombinant SAK has been produced in bacteria such as *Escherichia coli* and shown to induce fibrin specific clot lysis in human plasma milieu in vitro (Matsuo et al. 1990). Attempts have been made to compare the fibrinolytic properties of SAK and streptokinase using animal models of venous thrombosis and other methods (Collen et al. 1992).

Another activator in clinical use is the acylated plasminogen–streptokinase activator complex (APSAC), in which human plasminogen with an acylated active site is complexed with bacterial streptokinase. APSAC has an extended therapeutically effective half-life in circulation relative to streptokinase (Smith et al. 1981).

Because of the drawbacks of the available plasminogen activators, attempts are underway to develop improved recombinant variants of these compounds. Despite significant limitations, streptokinase remains the drug of choice, particularly in poorer countries, because of its low cost (Adams et al. 1991; Nicolini et al. 1992; Marder 1993; Wu et al. 1998).

Extracellular streptokinase is produced by various strains of β-hemolytic streptococci. The enzyme is a single-chain polypeptide that exerts its fibrinolytic action indirectly by activating circulatory plasminogen. The complete amino acid sequence of streptokinase was first established by Jackson and Tang (1982). Streptokinase has molecular weight of 47 kDa and is made up of 414 amino acid residues. The protein exhibits its maximum activity at pH of approximately 7.5, and its isoelectric pH is 4.7 (Taylor and Botts 1968; Brockway and Castellino 1974). The protein does not contain cystine, cysteine, phosphorus, conjugated carbohydrates, or lipids. Other physical and chemical data on streptokinase have been reported by De Renzo et al. (1967) and Taylor and Botts (1968). Streptokinases produced by different groups of streptococci differ considerably in structure (Huang et al. 1989; Malke 1993).

2.1 Structure and Mechanism of Action

A considerable degree of heterogeneity exists among the streptokinases produced by different groups of streptococci. Biophysical techniques such as circular dichroism, nuclear magnetic resonance (NMR) spectroscopy, Fourier-transform infrared (FT-IR) spectroscopy, and differential scanning calorimetry (DSC) have provided most of the available structural information on streptokinase (Beldarrain et al. 2001). Studies on fragments of streptokinase have also yielded valuable information (Shi et al. 1994).

Streptokinase consists of multiple structural domains (α-, β-, and γ-domains) with different associated functional properties (Fig. 3). Scanning calorimetric analysis suggests that the protein is composed of two distinct domains. The N-terminal domain (residues 1–59) has been found to complement the low plasminogen activation ability of the 60–414 amino acid residue domain of the protein (Nihalani et al. 1998).

Fig. 3 Crystal structure of
streptokinase β-domain

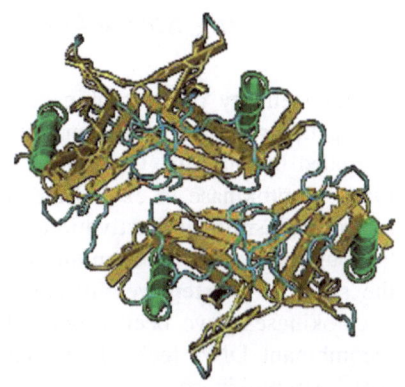

Recently, the crystal structure of streptokinase complexed with human plasmin light chain has been studied (Wang et al. 1998). Similarly, the crystal structure of streptokinase β-domain has been discussed (Wang et al. 1999). Thermal denaturation of streptokinase has been discussed (Azuaga et al. 2002), and the heat stability of specific domains of the enzyme has been studied (Conejero-Lara et al. 1996).

How streptokinase activates plasminogen has been the focus of extensive research (Zhai et al. 2003); nevertheless, mechanisms of activation of plasminogen by streptokinase are still being elucidated (Boxrud et al. 2001).

Streptokinase is known to activate plasminogen by both fibrin-dependent and fibrin-independent mechanisms (Reed et al. 1998). Streptokinase interacts with plasminogen though multiple domains. At least two independent plasminogen binding sites of streptokinase have been identified (Nihalani and Sahni 1995). The C-terminal 3 domain of streptokinase is involved in plasminogen substrate recognition and activation (Zhai et al. 2003). Similarly, the Asp[41]–His[48] region of streptokinase is important in binding to the substrate plasminogen (Kim et al. 2000). The role of an adjacent region, i.e., residues 48–59, on plasminogen activation has been discussed (Wakeham et al. 2002). The coiled region of the streptokinase γ-domain is said to be essential for plasminogen activation (Wu et al. 2001). Similarly, the streptokinase β-domain is involved in forming the streptokinase–plasminogen complex that is responsible for activating the plasminogen (Robinson et al. 2000).

Streptokinase binds preferentially to the extended conformation of plasminogen through the lysine binding site to trigger conformational activation of plasminogen (Boxrud and Bock 2000; Boxrud et al. 2001). Streptokinase–plasminogen activator complex interacts with plasminogen through long-range protein–protein interactions to maximize catalytic turnover (Sundram et al. 2003). The first 59 amino acid residues seem to have multiple functional roles in streptokinase (Shi et al. 1994). Without these N-terminal residues, streptokinase has an unstable secondary structure. Loss of residues 1–59 greatly reduces the activity of the remaining streptokinase fragment (i.e., residues 60–414; Young et al. 1995). Species-specific plasminogen activation has been demonstrated using 56 isolates of the pathogenic group C streptococci (McCoy et al. 1991).

2.2 Enhancing Streptokinase

Immunogenicity of streptokinase and its short half-life in circulation limit the therapeutic potential of this enzyme. Streptokinase in circulation is proteolytically degraded by plasmin. Consequently, research has focused on structurally modifying streptokinase to extend half-life, reduce or eliminate immunogenicity, and improve plasminogen activation. Any structural change needs to be informed by thorough structure–function analysis of streptokinase domains, and this has been the subject of extensive investigation (Zhai et al. 2003). Structurally modified streptokinases have been produced in several ways including genetic mutation, recombinant DNA technology, and chemical or enzymatic modification of the native streptokinase.

Mutant streptokinase with improved stability has been prepared (Shi et al. 1998). Two of the major sites of the proteolytic action of plasmin on streptokinase are Lys^{59} and Lys^{386} (Wu et al. 1998). This knowledge has been used to engineer variants of streptokinase that are resistant to plasmin (Wu et al. 1998) and, therefore, have a longer functional half-life. The plasmin-resistant forms of streptokinase have been found to be as active as the native streptokinase. Recombinant streptokinase produced in the yeast *Pichia pastoris* is glycosylated, and this appears to enhance its resistance to proteolysis (Pratap et al. 1996). Following a similar path, attempts have been made to extend the half-life of native nonglycosylated streptokinase by complexing it with polymers such as polyethylene glycol (PEG; Koide et al. 1982). Plasmin-resistant, long-life variants of protein-engineered streptokinase have been produced in a protease-deficient recombinant *Bacillus subtilis* WB600 (Wu et al. 1998). It appears that the streptokinase domains responsible for activity, stability, and immunogenicity have considerable overlap. Therefore, a future therapeutically optimal streptokinase will not necessarily be the most active nor the longest lived.

The immunogenicity of streptokinase became known soon after its discovery (Tillett and Garner 1933). Early studies showed that blood from patients with recent streptococcal infection inactivated streptokinase because of the presence neutralizing antibodies. Because streptococcal infections are common, a detectable level of antibodies against streptokinase occurs in most populations (Kazmi et al. 2002). Many different antistreptokinase platelet-activating antibodies are known to exist and occur widely (Regnault et al. 2003).

The different domains of streptokinase differ in immunogenicity (Reed et al. 1993). Extensive work has been reported on the identification of antigenic regions of streptokinases (Coffey et al. 2001). New antigenic domains were identified as recently as 2001 (Coffey et al. 2001).

Recombinant streptokinases with reduced immunogenicity have been produced (Ojalvo et al. 1999). A mutant streptokinase that lacked the C-terminal 42 amino acids was found to be less immunogenic than the native molecule (Torrens et al. 1999). One chemical modification has involved complexing streptokinase with PEG (Pautov et al. 1990), primarily for reducing immunogenicity.

Streptokinase variants with one or more of the normal amino acids residues replaced by others have been prepared in attempts to enhance plasminogen activation (Wu and Thiagarajan 1996). Some of the modified variants displayed enhanced stability. The preferred variant had Lys[59] replaced with a Glu-X residue. Streptokinase derivatives having platelet glycoprotein-binding domains have been reported (Galler 2000). These derivatives produced higher local concentrations of plasmin in vivo when compared with unmodified streptokinase.

2.3 Production of Streptokinase

2.3.1 Producing Microorganisms

Streptokinase-producing streptococci were first identified by Billroth (1874) in exudates of infected wounds. Later, the blood of scarlet fever patients was shown to contain similar microorganisms. By 1919, *Streptococcus* sp. had been classified into α, β, and γ variants based on the distinct types of hemolytic reactions that the variants produced on blood agar plates.

In 1933, Lancefield used serologic distinctions to further differentiate the β-hemolytic streptococci into groups A to O (Lancefield 1933). Most of the streptokinases are obtained from β-hemolytic streptococci of the Lancefield groups A, C, and G. The group C are preferred for producing streptokinase as they lack erythrogenic toxins. The group C strain *Streptococcus equisimilis* H46A (ATCC 12449) isolated from a human source in 1945 has been widely used for producing streptokinase. The strain H46A was selected from more than 100 fibrinolytic isolates, because it yielded the most active streptokinase.

S. equisimilis H46A does not produce erythrogenic toxins and is less fastidious in its growth requirements than the majority of group A strains (Christensen 1945). The H46A isolate can be grown on semisynthetic media to secrete large quantities of streptokinase (Feldman 1974). H46A is also the main source of the streptokinase gene that has been expressed in various other microorganisms (Wong et al. 1994).

2.3.2 Recombinant Producers

The streptokinase gene from *S. equisimilis* H46A was sequenced by Malke et al. (1985). The transcriptional control of this gene has been studied (Gase et al. 1995), and functional analysis of its complex promoter has been reported (Grafe et al. 1996). Considerable information exists, therefore, for effective use of this gene in producing streptokinase safely in nonpathogenic bacteria.

Studies of the streptokinase gene isolated from various sources suggest it to be polymorphic (Malke 1993). The streptokinase gene (*skc*) cloned from *S. equisimilis* H46A has been expressed in several Gram-positive and Gram-negative bacteria including *B. subtilis* WB600 (Wu et al. 1998) and *E. coli* (Yazdani and Mukherjee 2002). Production in the yeast *Pichia* has been reported (Pratap et al. 2000).

Plasmids bearing the *skc* and erythromycin resistance genes have been introduced into *S. equisimilis* H46A in attempts to select overproducer clones (Muller and Malke 1990). Plasmids designed for high-level secretory expression of streptokinase have been successfully evaluated for producing it (Ko et al. 1995). An ability to produce recombinant streptokinase greatly enhances the possibilities for beneficial structural modifications of this protein and enhanced production of the desired recombinant streptokinase.

2.3.3 Fermentation Process

Group A hemolytic streptococci are highly fastidious, commonly requiring complex and rich media supplemented with various nutritional factors for growth (Bernheimer et al. 1942). A modification of the medium originally developed by Bernheimer et al. (1942) was used by Christensen (1945) for producing streptokinase from *S. equisimilis* H46A. This medium contained peptone, phosphate salts, glucose, biotin, riboflavin, tryptophan, glutamine, and nucleic acids (thiamine, adenine, and uracil). The glutamine content was only 25% of that in the original medium of Bernheimer et al. (1942). Low glucose concentration at inoculation allowed growth to become well-established without excessive production of acid from sugar fermentation. Further glucose was added after the initial overnight incubation. Cells proliferated to high density. Alkali (5 M sodium hydroxide) was added at regular intervals to maintain neutral pH. In the absence of pH control, both biomass and streptokinase production were adversely affected.

Batch cultures of streptococci have been characterized as having a rather extended lag phase followed by a relatively short period of exponential growth due to nutrient depletion. Comparison with continuous culture has revealed lower productivity of the equivalent batch fermentation. Studies on glucose and tryptophan limitation in continuous cultures of *Streptococcus faecalis* concluded that maximum biomass yield per unit energy source was attained when the energy source (glucose) was limiting. In contrast, production of streptokinase was maximized when glucose was in excess and the other nutrients were present in limiting amounts (Rosenberger and Elsden 1960).

Addition of polyoxyethylene sorbitan monooleate at concentration of 0.01–0.10% to the medium prior to inoculation has been recommended for achieving high yield of streptokinase. Effect of amino acid feeding on steady-state continuous cultures of group A streptococci was reported by Davis et al. (1965). The culture had a dilution rate of 0.5 h^{-1} at a constant pH value of 7.5. Of the 14 amino acids tested, all except alanine could limit growth.

A study with continuous cultures of a group C β-hemolytic *Streptococcus* strain H46 assessed the biomass and streptokinase production at various dilution rates, pHs, and temperatures in a complex medium with excess glucose. At pH 7.0, the yield of biomass and streptokinase on glucose did not vary with the dilution rate;

however, the biomass and streptokinase productivities increased with increasing dilution rate in the range of 0.1–0.5 h^{-1}. Continuous culture had more than twice the streptokinase productivity of batch culture (Holmstrom 1965).

In a study of 19 strains of group E streptococci (SGE), 18 of the isolates were found to produce streptokinase in an enriched tryptose broth medium (Ellis and Armstrong 1971). This medium contained tryptose, glucose, sodium chloride, uracil, adenine, glutamine, tryptophan, vitamins, and salts. A medium containing corn steep liquor, cerelose, KH_2PO_4, and $KHCO_3$, pH 7.0, was used by Feldman (1974) for producing streptokinase. Use of corn steep liquor instead of casein hydrolyzate enhanced streptokinase yield in cultures of the strain H46A. Baewald et al. (1975) used a simple and inexpensive medium to obtain high yields of streptokinase from *S. equisimilis*. The medium contained yeast autolyzate or corn steep liquor as the nitrogen source, glucose, and various salts. High titers of streptokinase were attained at 28°C, pH 7.2–7.4, within 24 h in agitated cultures.

A commonly used complex medium for producing streptokinase is brain–heart infusion (BHI) medium. This medium is produced using bovine brain and heart tissue. Malke and Ferretti (1984) obtained good proliferation of *S. equisimilis* H46A on BHI medium. The same medium was used by Suh et al. (1984) to produce streptokinase from a group C *Streptococcus* isolated from a human patient. The maximum activity of streptokinase was detected during exponential growth. A glucose and casein hydrolyzate medium supplemented with various salts was reported by Nemirovich-Danchenko et al. (1985) for producing strep-tokinase from *S. equisimilis*.

Chemically defined media (CDM) for growing group A streptococci have been developed to require only small inocula and without the need for a prior adaptation regimen. The doubling times of the streptococci in such media can be comparable to those in complex media. Group A streptococci grown in defined media can produce streptokinase and other secreted enzymes (McCoy et al. 1991).

Influence of cultivation temperature on growth and streptokinase production of *S. equisimilis* H46A has been investigated over the temperature range of 28–43°C. The H46A strain was capable of growth over the entire temperature range, but 28°C was the optimal growth temperature (Ozegowski et al. 1983).

Relatively recently, work has focused on elucidating the fermentation conditions for producing streptokinase from mutants of the wild-type streptococci and other genetically engineered microorganisms. Hyun et al. (1997) produced copious quantities of streptokinase using a mutant *Streptococcus*. The culture medium consisted of casein or serration peptidase hydrolyzed casein, glutamine, cysteine, and yeast extract. The mutant was cultured at pH 6.8–7.2, 35–38°C, in broth aerated at 0.1–1.0 rpm. The titer of streptokinase exceeded 8,500 U/ml.

The *S. equisimilis* streptokinase gene expressed in *E. coli* led to a 10-fold greater streptokinase titer than values obtained in cultures of group C streptococci such as *S. equisimilis* (Estrada et al. 1992). Work has been reported on localizing the core promoter region of the streptokinase gene, *skn* (Malke et al. 2000). The plasmid pSK100 has enabled high-level secretory expression of streptokinase in *E. coli* cultured in ampicillin-containing LB medium. Expression titers of about

5,000 U/ml have been attained (Ko et al. 1995). LB medium consists of Bactot-ryptone, yeast extract, sodium chloride, and 50 μg ml^{-1} ampicillin at pH 7.3. Other media have been described for producing recombinant streptokinase in *E. coli* (Narciandi et al. 1996).

A recombinant streptokinase was produced in *E. coli* using LB medium at 37°C (Lee et al. 1997a). The plasmid used imparted ampicillin resistance to the bacterium, and ampicillin provided the selection pressure for plasmid retention. The production of streptokinase was induced by adding 1 mM isopropyl-*β*-*D*-thiogalactopyranoside (IPTG) to the medium. The plasmid encoded a streptokinase that lacked the 13 *N*-terminal amino acid residues of the normal protein. This enhanced productivity of the recombinant protein and enabled secretion into the extracellular medium (Lee et al. 1997a, b). At least a part of the N-terminal domain is known to be functionally relevant in streptokinase (Mundada et al. 2003), but this may not include the first 13 residues (Lee et al. 1997b). Other reports have also described IPTG-induced production of recombinant streptokinase (Yazdani and Mukherjee 1998). The specific productivity of the recombinant protein was substantially enhanced by fedbatch cultivation compared with batch fermentations of IPTG-induced *E. coli* (Yazdani and Mukherjee 1998). Some work on kinetic analysis and modeling of streptokinase fermentation has been reported (Stuebner et al. 1991).

2.4 Assaying Streptokinase

Characterization of streptokinase production by fermentation and assessment of the product require methods for assaying the streptokinase. Unlike for enzymes such as *β*-galactosidase, no synthetic substrate has yet been identified for the streptokinase assay. Plasminogens of human, chimpanzee, monkey, cat, dog, and rabbit are the only known protein substrates for streptokinase (Castellino 1981). Streptokinase assays rely on its ability to activate plasminogen to plasmin. Plasmin then hydrolyzes an indicator substrate, and the extent of hydrolysis over a given period is related back to the concentration of streptokinase. The indicator substrates for plasmin include the fibrin clot, casein, other proteins and various synthetic esters (lysine methyl ester, lysine ethyl ester, *p*-toluenesulfonyl-L-arginine methyl ester).

In 1949, Christensen devised the first quantitative method for determining streptokinase (Christensen 1949). This procedure required the determination of the smallest quantity of the streptokinase-containing solution to cause lysis of a standard fibrin clot in 10 min at 37°C, pH 7.4. Various modifications of this procedure have been used widely (Tomar 1968). Digestion of casein for estimating streptokinase was established as early as 1947. A radial caseinolysis method with agarose gel containing both casein and plasminogen is commonly used (Saksela 1981). A quantitative enzymatic assay for human plasminogen and plasmin using azocasein as substrate was described by Hummel et al. (1965).

The fibrin plate method originally introduced for determining proteolytic activity in blood has been widely used for measuring fibrinolysis. The method has been criticized for various shortcomings, and numerous attempts have been made to improve it. In the fibrin plate method, a zone of lysis produced on a fibrin film in a Petri dish is measured and related to the concentration of the fibrinolytic enzyme (Permin 1947).

A heated fibrin plate method was advanced by Lassen (Lassen 1952). An agar diffusion assay for streptokinase was devised by Holmstrom (1965) as a modification of the Astrup and Mullertz method. A rapid fibrin plate assay was designed by Marsh and Gaffney (1977). This procedure reduced the required incubation time to 3 h from the 16–20 h needed for the conventional fibrin plate assay. Incubation period was reduced by using fibrin films that had been enriched with plasminogen. Studies of fibrin clot lysis by streptokinase have identified pH, buffer concentration, and plasminogen concentration as the major variables influencing the lysis (Westlund and Andersson 1991). The concentration of fibrinogen in the clot also affects the plate assay.

Synthetic amino acid esters were first introduced as sensitive substrates for assaying proteolytic enzymes (Mullertz 1954) and soon afterwards were shown to be suitable for assaying plasmin (Troll et al. 1954; Sherry et al. 1964). Streptokinase–plasminogen/streptokinase–plasmin complex hydrolyzes methyl esters of lysine, acetyl-lysine, and tosyl-arginine (Reddy et al. 1974). Hydrolysis of synthetic esters by plasmin generated by activating plasminogen with streptokinase can be followed by determination of the carboxylic acid functional groups produced. Hestrin's photometric method established in 1949 can be used for this determination. This method relies on the formation of a ferric complex with hydroxamic acid (Mullertz 1954). Alternatively, the free carboxylic acid groups produced by hydrolysis of synthetic esters can be determined by titrimetry (Davie and Neurath 1953).

A modified Hestrin method has also been described (Roberts 1958). Use of synthetic esters instead of fibrinogen has enabled a consistent and rigorous definition to be established for the specific activity of streptokinase (Taylor and Botts 1968). The use of Na-CBZ-lysine-p-nitrophenyl ester (CLNE) as a substrate for plasmin has been reported. The assay based on this substrate is substantially more specific and convenient compared with assays that use Na-tosyl-L-arginine methyl ester (TAME) or Na-benzoyl-L-arginine ethyl ester (BAEE; Silverstein 1975). Compared with the TAME-based assay, use of CLNE is 800-fold more specific for streptokinase–plasminogen/streptokinase–plasmin (SK-PG/SK-PN) complex. Similarly, the CLNE substrate is 40-fold more specific for SK-PG/SK-PN complex than is BAEE. A colorimetric plasmin assay based on TAME was developed by Castellino et al. (1976).

A solid-phase chromogenic assay for plasmin has been reported (Kulisek et al. 1989). In this assay, the plasmin generated by the action of nitrocellulose-bound streptokinase hydrolyzes the plasmin-specific tripeptide H-D-valyl-leucyl-lysine-p-nitroaniline. Other similar spectrophotometric methods for assaying streptokinase have been described.

A highly sensitive and reproducible enzyme-linked immunosorbent assay (ELISA) for antibodies against streptokinase has been developed (Leonardi et al. 1983). An enzyme immunoassay for streptokinase was reported by Shemanova et al. (1995). This assay was 10 times more sensitive than the other methods used typically and required only microquantities of blood serum (cf. ≥ 2 ml of plasma is required for a typical plate or test-tube assay). Eisenberg et al. (1990) reported the use of ELISA for measuring the cross-linked fibrin degradation products (FDPs) in plasma. For more details see chapter clot lysis tests.

2.5 Recovery and Purification

Several schemes have been described for recovery and purification of streptokinase either from the commercially available crude preparations or the fermentation broths of various streptococci (Perez et al. 1998).

De Renzo et al. (1967) purified streptokinase from a relatively crude commercial preparation (Varidase; Lederle Laboratories, American Cyanamid, USA). Column chromatography on DEAE-cellulose was followed by column electrophoresis in sucrose density gradients to obtain a five- to sixfold increase in purity. Repeated chromatography was necessary to remove the last detectable traces of impurities. In a similar procedure, starting from crude Varidase, Taylor and Botts (1968) attained a final specific activity of 100,000 units of streptokinase per mg of protein. This required a combination of ion exchange (DEAE-Sephadex A-50) and gel permeation (Sephadex G-100) chromatography.

Tomar (1968) purified streptokinase from Varidase using a different procedure. Streptokinase was fractionated either by hydroxyapatite chromatography or ammonium sulfate fractionation. Precipitation of streptokinase with 40–50% ammonium sulfate resulted in a two- to threefold increase in specific activity. The precipitate was recovered by centrifugation and dialyzed against 0.09 M sodium chloride. The dialyzed solution was further purified by gradient elution from a DEAE-cellulose chromatography column. The major peak of eluted activity was concentrated 10-fold by ultrafiltration.

A highly pure streptokinase was recovered from the relatively crude commercial Kabikinase (Kabi Vitram, Sweden) by Einarsson et al. (1979). Ammonium sulfate fractionation was first used to obtain a crude precipitate of streptokinase. This was redissolved and subjected to gel permeation chromatography. The eluted streptokinase fraction was further purified using column chromatography on DEAE-cellulose or DEAE-Sepharose.

Several affinity chromatography methods have been discussed for purifying streptokinase (Rodriguez et al. 1994). The earliest such procedure used insolubilized di-isopropyl fluorophosphate (DIP) plasmin as the affinity ligand. This ligand was produced by the conversion of plasminogen to plasmin with urokinase and inhibition of the proteolytic activity of plasmin by DIP (Castellino et al. 1976). Purification through the affinity column caused a 30% decrease in the streptokinase activity, suggesting incomplete inhibition of the plasmin affinity ligand bound to the chromatography matrix. A different affinity purification used a monoclonal antibody ligand (Andreas 1990).

Another affinity purification scheme used acylated plasminogen or plasmin as the affinity ligand (Rodriguez et al. 1992). The acylation of plasminogen or plasmin was carried out with p-nitrophenyl-p-guanidino benzoate (NPGB). Plasminogen acylation with NPGB allowed it to be used as an affinity ligand without requiring activation to plasmin. This probably reduced the plasmin-associated proteolysis of streptokinase. Rodriguez et al. (1994) used a combination of two affinity matrices for chromatographic purification of recombinant streptokinase. The affinity ligands were human plasminogen and monoclonal antibody against streptokinase. Both ligands were bound to Sepharose as the chromatographic matrix. This purification method produced a preparation with about 50,000 units of activity per mg of protein and purity >93%.

Use of immobilized NPGB-acylated plasminogen for affinity purification was further reported by Liu et al. (1999). A solution of urea was the eluent. Approximately ninefold purification was achieved with yield >90%. The specific activity of the purified material was 74,000 units/mg. Hernandez-Pinzon et al. (1997) recovered streptokinase by cross-flow ultrafiltration. Approximately 14% of the initial activity was lost as the protein solution was concentrated by eightfold. This loss was associated with denaturation of the streptokinase through unfolding and aggregation.

Streptokinase has been purified from the filtrate of a streptococcal fermentation broth using hydrophobic interaction chromatography on phenyl- or octyl-Sepharose column. A gradient elution with 21% ammonium sulfate was used to recover the streptokinase. Further purification involved gel permeation and ion-exchange chromatography steps.

Perez et al. (1998) purified a recombinant streptokinase produced by fermentation in $E.$ $coli$ K12. To isolate the streptokinase, the biomass was first recovered by centrifugation and then washed, and the cells were disrupted. The streptokinase inclusion bodies were then dissolved and renatured. Hydrophobic interaction chromatography was then used to obtain the protein at purity >99%. The overall recovery yield was 49%. A similar yield of about 45% and purity >97% were attained by Zhang et al. (1999) in recovering recombinant streptokinase from inclusion bodies produced in $E.$ $coli$. Generalized schemes for recovery and renaturation of inclusion body recombinant proteins have been published (Chisti 1998a).

Reverse-phase high-performance liquid chromatography (HPLC) has been used for purifying a bovine plasminogen activator from culture supernatants of the bovine pathogen $S.$ $uberis$. A single protein with molecular mass of 32 kDa was detected in the eluted active fraction. This plasminogen activator lacked the C-terminal domain that is characteristic of the $S.$ $equisimilis$ streptokinase. Other work on $S.$ $uberis$ streptokinase has been reported (Johnsen et al. 1999, 2000), and its mechanism of action has been subjected to closer scrutiny (Sazonova et al. 2000).

Commercial production of streptokinase requires special attention to biosafety considerations, because the protein is potentially immunogenic to process workers. In addition, care is necessary if streptokinase is being produced using natural strains of streptococci, because all streptokinase-producing streptococci are potentially pathogenic. The various safety considerations relevant to production of biopharmaceutical proteins have been discussed by Chisti (1998b).

3 Staphylokinase

3.1 Structure of Staphylokinase

Staphylokinase is a single polypeptide chain of 136 amino acids without disulfide bridges that is secreted by certain strains of *Staphylococcus aureus*. Like streptokinase, staphylokinase is not an enzyme, but it forms a 1:1 stoichiometric complex with plasmin(ogen) that activates other plasminogen molecules. The structure of the ternary plasmin–staphylokinase–plasminogen complex (Fig. 4) has recently been resolved by computer modeling and scanning mutagenesis (Jespers et al. 1998, 1999).

3.2 Mechanism of Fibrin Selectivity

When staphylokinase is added to human plasma containing a fibrin clot, it reacts poorly with plasminogen, but reacts with high affinity with traces of plasmin at the clot surface. At the clot surface, the plasmin–staphylokinase complex efficiently activates plasminogen to plasmin. Both plasmin staphylokinase and uncomplexed plasmin bound to fibrin are protected from rapid inhibition by α_2-antiplasmin, whereas their unbound counterparts, liberated from the clot or generated in plasma, are rapidly inhibited by α_2-antiplasmin. Thereby, the process of plasminogen activation is confined to the thrombus, preventing excessive plasmin generation, α_2-antiplasmin depletion, and fibrinogen degradation in plasma. The biochemical pathways governing these fibrin-selective interactions were summarized by Collen (1998a, b).

3.3 Pharmacokinetics and Thrombolytic Properties in Patients

In patients with acute myocardial infarction treated with intravenous infusion of 10 mg staphylokinase over 30 min, staphylokinase-related antigen disappeared from plasma in a biphasic mode with a $t_{1/2\alpha}$ of 6.3 min and $t_{1/2\beta}$ of 37 min, corresponding to plasma clearance of 270 ml/min (Collen and Van de Werf 1993). This short initial half-life and rapid clearance would seem to predestine staphylokinase for administration by continuous infusion or double bolus injection (see below). We have recently been able, however, to reduce the clearance 5-fold to 30-fold by selective chemical substitution with single PEG molecules with M_r of 5,000–20,000, as demonstrated in experimental animal models and in pilot studies in patients with acute myocardial infarction.

Recombinant staphylokinase was compared with accelerated, weight-adjusted alteplase in two open, randomized studies, each involving 100 patients with acute

Fig. 4 Crystal structure of the ternary complex between plasmin (microplasmin, μPli, *green*), staphylokinase (SAK, *blue*), and plasminogen (microplasminogen, μPli, *red*). The molecules are depicted in *arrowed ribbon representation*, and the side-chains of the catalytic triad residues in μPli are represented as *dark-green sticks*. The μPlg substrate molecule is docked in the active site cleft and covers 1,140 Å2 and 1,190 Å2 contact surface with SAK and μPli, respectively (Jespers et al. 1998, 1999)

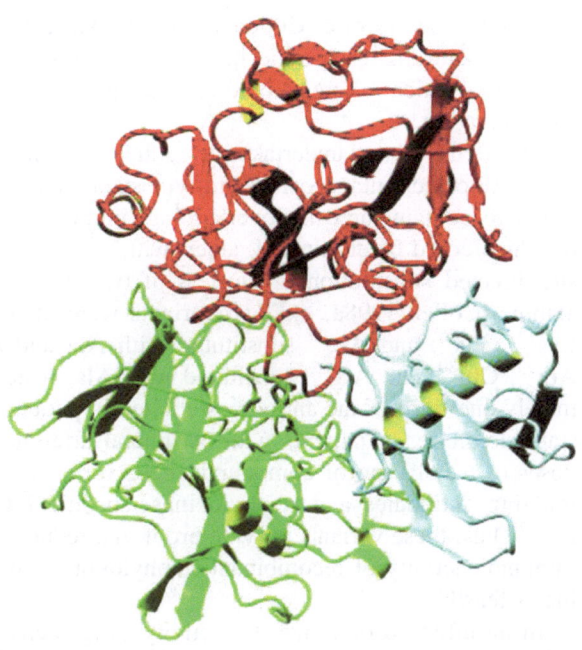

myocardial infarction (Collen 1998a, b). In both studies, recombinant staphylo-kinase was found to be at least equipotent to alteplase in terms of complete arterial recanalization within 90 min. Staphylokinase was highly fibrin selective, as revealed by virtually unaltered levels of plasma fibrinogen, plasminogen, and α_2-antiplasmin. No strokes, allergic reactions, or other side-effects were recorded. Thus, intravenous staphylokinase, combined with heparin and aspirin, is a potent, rapidly acting, and highly fibrin-selective thrombolytic agent in patients with acute myocardial infarction.

Recombinant staphylokinase variants have been infused intra-arterially as a 2 mg bolus followed by 1 mg/h infusion in 178 patients with angiographically confirmed peripheral arterial occlusion. Recanalization within 48 h was complete in 148 patients (83%), partial in 23 (13%), and absent in 7 (4%) patients. Major hemorrhagic stroke occurred in three patients and was fatal in two. Blood trans-fusion was required in 15 patients (8.4%). One-year follow-up in 161 patients revealed amputation-free survival in 138 (86%).

3.4 Immunogenicity

Levels of preformed antistaphylokinase antibodies in the general population are lower than those of antistreptokinase antibodies, whereas the current clinical experience in over 300 patients suggests that major allergic reactions to staphylokinase are rare (Collen 1998a, b). Most patients, however, develop high titers of neutralizing specific immunoglobulin G (IgG) after infusion of staphylokinase, which would predict therapeutic refractoriness upon repeated administration.

Efforts have been undertaken to reduce the immunogenicity of staphylokinase by site-directed mutagenesis. Wild-type staphylokinase (SakSTAR variant 36) was found to contain three nonoverlapping, immunodominant epitopes, at least two of which could be eliminated, albeit with partial inactivation of the molecule, by site-directed substitution of clusters of two or three charged amino acids with alanine (Collen 1998a, b). Two variants were identified, one with Lys^{35}, Glu^{38}, Lys^{74}, Glu^{75}, and Arg^{77} substituted with Ala and the other with Lys^{74}, Glu^{75}, Arg^{77}, Glu^{80}, and Arg^{82} substituted with Ala, which did not recognize approximately one-third of the antibodies elicited in patients by treatment with wild-type staphylokinase. In patients with peripheral arterial occlusion given intra-arterial doses of 6.5–12 mg of compound, these variants induced significantly less neutralizing antibodies and staphylokinase-specific IgG than wild-type staphylokinase. Thus, these variants provide proof that reduction of the immunogenicity and immunoreactivity of recombinant staphylokinase in humans by protein engineering is feasible.

In an effort to optimize the activity/antigenicity ratio, a comprehensive site-directed mutagenesis study was carried out. Over 350 plasmid-encoding staphylokinase mutants were constructed and expressed in *E. coli*, and the expression products were purified and characterized. Comprehensive analysis of combination variants led to the identification of staphylokinase (E65D, K74R, E80A, D82A, K130T, K135R) (Birkedal-Hansen 1995) with intact, specific activity and that bound less than 50% of the antibodies of pooled plasma of 40 patients treated with wild-type staphylokinase. Intra-arterial administration of this variant in 18 patients with peripheral arterial occlusion induced complete recanalization in 16, but significantly less circulating neutralizing antibodies after 3–4 weeks than with staphylokinase. Overt neutralizing antibody induction (>5 µg compound neutralized per milliliter plasma) was observed in 56 of the 70 patients (80%) given wild-type staphylokinase, but in only 5 of the 18 patients treated with staphylokinase (E65D, K74R, E80A, D82A, K130T, K135R). In a final round of mutagenesis, approximately 100 additional plasmids were constructed, expressed, purified, and characterized, yielding staphylokinase (K35A, E65Q, K74Q, D82A, S84A, T90A, E99D, T101S, E108A, K109A, K130T, K135R, K136A, ▼137) with maintained fibrinolytic potency and fibrin selectivity in human plasma milieu and markedly reduced reactivity with antistaphylokinase antibodies in pooled, immunized patient plasma. Intra-arterial administration in patients with peripheral arterial occlusion induced SakSTAR-neutralizing activity, exceeding 5 µg/ml plasma in two of seven patients. Thus, staphylokinase variants with markedly reduced antibody induction but intact thrombolytic potency can be generated.

3.5 PEG-Derivatized Cysteine-Substitution Variants for Single Bolus Administration

Bolus administration of thrombolytic agents is becoming a preferred regimen for thrombolytic therapy of acute myocardial infarction. Derivation of proteins with PEG may reduce their clearance while maintaining their specific activity. Therefore, a recombinant staphylokinase variant with reduced immunogenicity in which Ser in position 3 of the protein sequence was mutated into Cys, namely staphylokinase (S3C, K35A, E65Q, K74R, E80A, D82A, T90A, E99D, T101S, E108A, K109A, K130T, K135R), was derivatized with maleimide-substituted PEG (P) with molecular weight of 5,000 (P5), 10,000 (P10), or 20,000 (P20), and characterized in vitro and in vivo.

The homogeneous 1:1 stoichiometric Cys-linked PEGylated variants had intact specific activities (140–180 kU/mg) and fibrin-selective thrombolytic potencies in human plasma milieu in vitro (50% clot lysis with 0.32–0.40 µg/ml, as compared with 0.24 µg/ml for wild-type staphylokinase), and their thermostability was maintained after 5 days at 37°C. PEGylation reduced the plasma clearance of an intravenous bolus in hamsters and rabbits approximately 5-fold with P5, 10-fold with P10, and 50-fold with P20. In hamsters, bolus injection induced dose-related lysis of a 50 µl ^{125}I-fibrin-labeled plasma clot injected in the jugular vein: 50% clot lysis at 90 min was obtained with 17, 15, and 8 µg/kg, as compared with 45 µg/kg with un-PEGylated staphylokinase.

Intravenous bolus injection of 5 mg of the PEGylated variants in nine patients with acute myocardial infarction restored thrombolysis in myocardial infarction-3 (TIMI-3) flow at 60 min in six patients. SakSTAR-related antigen disappeared from plasma with an initial half-life of 15, 30, and 120 min and was cleared at a rate of 70, 40, and 8 ml/min for variants substituted with P5, P10, and P20, respectively, as compared with an initial half-life of 3 min and a clearance of 360 ml/min for wild-type staphylokinase. On the basis of these results, staphylokinase (S3C, K35A, E65Q, K74R, E80A, D82A, T90A, E99D, T101S, E108A, K109A, K130T, K135R) substituted with a single PEG molecule with molecular weight of 5,000 linked to a cysteine residue introduced in position 3 of the amino acid sequence has been selected for clinical development as a single intravenous bolus agent for thrombolytic therapy of acute myocardial infarction.

4 Serrapeptase

Serrapeptase, also known as serratiapeptase, serratiopeptidase, Serratia E-15 protease, serralysin, serratia peptidase, serratio peptidase, or serrapeptidase (Fig. 5), is a proteolytic enzyme produced by enterobacterium *Serratia* sp. E-15 (Nakahama et al. 1986). This microorganism was originally isolated in the late 1960s from silkworm *Bombyx mori* L. (intestine) (Miyata et al. 1970). The true purpose of this

Fig. 5 Crystal structure of serrapeptase with co-ordinated zinc (*grey*) and calcium (*white*)

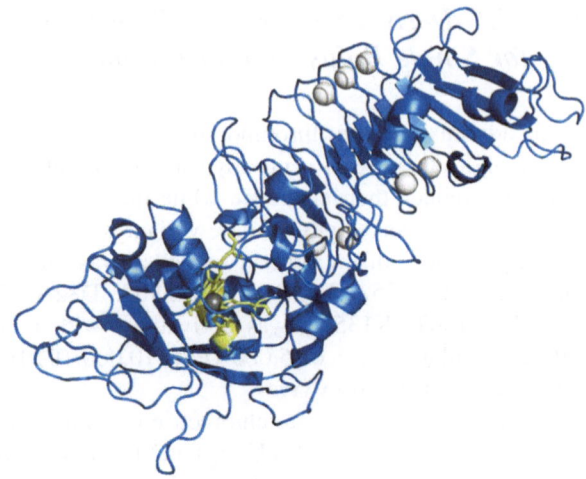

organism is to help silkworms dissolve their own cocoons. In addition, serrapeptase is adept at dissolving blood clots, arterial plaques, necrotic tissues, and inflammatory factors (Mazzone et al. 1990).

This enzyme has been used clinically in Europe and Asia for nearly a quarter century, for its anti-inflammatory actions to treat conditions, such as chronic sinusitis, sprains and strains, edema, and even postoperative inflammatory states, and for thinning of bronchopulmonary secretions. New research on this novel enzyme demonstrates its efficacy for treating several disease states. Studies on serrapeptase have focused on its use for treating chronic lung disease; ear, nose, and throat disorders; carpal tunnel syndrome; and edema following injury and surgery (Tachibana et al. 1984; Esch et al. 1989; Mazzone et al. 1990; Nakamura et al. 2003).

In patients with chronic airway disease (in which mucus production and removal are problematic), treatment with 30 mg/day serrapeptase for 4 weeks resulted in changes in sputum. Weight, viscosity, elasticity, and neutrophil content were all decreased. Coughing and expectoration frequency were significantly decreased (Nakamura et al. 2003). Using serrapeptase for treating chronic lung conditions in which sputum production is a problem (for example, cystic fibrosis) leads to improved lifestyle parameters.

How it Works?

Serratiopeptidase breaks down artherosclerotic plaques accumulated in arteries. Because the enzyme digests nonliving tissue and leaves live tissue alone, it may be effective in removing deposits of fatty substances, cholesterol, cellular waste products, calcium, and fibrin on the inside of arteries. The fibrinolytic (clot removal) activity of serratiopeptidase makes it an excellent blood-thinning and clot-dissolving agent. It is believed that serratiopeptidase acts upon inflammation by thinning the fluids in the body that collect around injured areas and increasing fluid drainage. This also enhances tissue repair and reduces pain. Pain is also reduced by the protein enzyme's ability to block amines. Serratiopeptidase also has the unique ability to dissolve the dead and damaged tissue that is a by-product of the healing response without harming living tissue. It is used in this way by the silkworm to digest a hole in the dead tissue of the cocoon so the silkworm can emerge. Serratiopeptidase also works by modifying cell-surface adhesion molecules, which guide inflammatory cells to their targets. These adhesion molecules are known to play an important role in the development of arthritis and other autoimmune diseases (Klein and Kullich 2000).

5 Nattokinase

5.1 History of Natto

After hours of fermentation, boiled soybeans metamorphose into an ancient medicinal food called *natto*. *Natto* may just be the perfect food, containing 18 valuable amino acids and an enzyme, nattokinase, that may challenge the pharmaceutical industry's best blood-clot busters. *Natto*, which has recently attracted attention throughout the world, is the third most popular type of fermented soybean in the Japanese diet. Japan has the highest average longevity in the world, which may partly be attributed to high consumption of *natto*.

When compared with ordinary soybeans, *natto* contains more calories, protein, fiber, calcium, potassium, and vitamin B2. Its high protein and economical price in terms of protein per gram has earned it the sobriquet "meat of the field." This nickname appears well deserved, as in comparison with an equivalent amount of beef, *natto* has slightly less protein (16.5–21.2 g) but contains more carbohydrates and fiber, and is also higher in calcium, phosphorus ions, and vitamin B2. Plus, it has nearly double the calcium and far more vitamin E to boot.

According to legend, the first person to originate traditional Japanese *natto* was the famous warrior Yoshiie Minamoto during the Heian Era of Japanese history (794–1192 AD). The horse was extremely important to the Japanese samurai warrior of the period, and great care was given to provide suitable provisions for the horses when armies were on the move. Typically, boiled soybeans were cooled down, dried in the sun, and packed immediately in rice straw bags for transport with the army. If the army was on a rapid deployment, the boiled soybeans were packed hastily into the rice straw bags without cooling or drying. The rice straw happened to contain a harmless and naturally occurring microorganism, *Bacillus subtilis*, that fermented the soybeans and produced *natto* with its characteristic sticky texture.

Initially, the soybeans were presumed to have spoiled, until Yoshiie Minamoto observed that his horses were picky eaters and yet demonstrated a preference for the spoiled soybeans or *natto*. One day, Minamoto demonstrated tremendous courage and dipped his finger into the seemingly rotten "goo". To his astonishment, the fermented soybeans were not only edible but had a distinct *umami* flavor. Minamoto was responsible for introducing *natto* to northwestern Japan, where he ruled. To this day *natto* is especially popular in that region of Japan and a folk remedy for fatigue, beriberi, dysentery, and heart and vascular diseases.

The most distinctive features of *natto* are the adhesive surrounding the soybeans and the strong flavor. The sticky material has been shown to consist of poly-g-glutamic acid (D and L) and polysaccharides, and the strong "cheese-like" flavor is due to the presence of pyrazine. These features sometimes make it hard for some people, especially people from other countries, to accept *natto*; however, these are the main factors which give *natto* its outstanding properties. *Natto*, which has recently attracted attention throughout the world, is a familiar part of the Japanese diet. Furthermore, it has recently been found that *natto* can help to prevent the viral infection O-157, osteoporosis, high blood pressure, and atherosclerosis, and dissolve fibrin (Sumi et al. 1987).

5.2 Discovery of Nattokinase

Sumi had long researched thrombolytic enzymes. He was searching for a natural agent that could successfully dissolved thrombius associated with cardiac and cerebral infarction (blood clots associated with heart attacks and stroke). One day in 1980, Dr. Sumi took the *natto* that he was eating for lunch and dropped a small portion into an artificial thrombus plate. The *natto* gradually dissolved the thrombus, and had completely resolved it in 18 h. Dr. Sumi named the corresponding fibrinolytic enzyme "nattokinase". Dr. Sumi commented that nattokinase showed a potency matched by no other enzyme.

In an experiment by Sumi et al. (1987) (Fig. 6), *natto* was applied directly to a fibrin plate. Nattokinase was extracted from 300 g *natto* with 220 ml saline for 15 min with stirring at 4°C. The material was filtered through gauze and then centrifuged at 3,000 rpm for 10 min. Nattokinase (10 nl, 21 mg protein/ml), urokinase standard 100 U/ml, and plasmin standard 4 CU/ml were applied to a fibrin plate. The nattokinase extract was heat-treated at precise temperatures for 10 min, and then a 10 μL sample was applied to a fibrin plate.

Dr. Sumi conducted research on about 170 kinds of food from all over world, and he found that *natto* had the highest fibrinolytic activity among all those foods. There are many traditional foods for prevention and treatment of thrombosis (e.g., azuki beans, Korean ginseng, and Japanese water dropwort), but most of these foods inhibit platelet aggregation similar to aspirin (antiplatelet). Only nattokinase acts only on the fibrinolytic system to dissolve within the blood vessels. Dr. Sumi presented the results of his research in Japan for the first time at the Japan Agricultural Chemistry Society. Later, he wrote a similar article for the International Thrombolytic Association, based on which the audience began to believe that dietary intake of *natto* was the major contributor to the longevity of Japanese people (Sumi et al. 1987, 1990, 1992).

Fig. 6 Historical discovery of nattokinase from *natto* in 1987 by Sumi and colleagues. In the experiment, *natto* was placed on a fibrin plate (**a**) and significantly lysed fibrin, as shown by the three halos. In (**b**) nattokinase extracted from *natto* was placed on the same fibrin plate next to urokinase (U) and plasmin (P). Nattokinase in vitro more than doubled the amount of fibrin lysis compared with urokinase and plasmin. (**c**) Nattokinase extract was heat-treated at high temperatures of 90°C, 80°C, and 70°C, yet remarkably continued to produce fibrinolytic activity

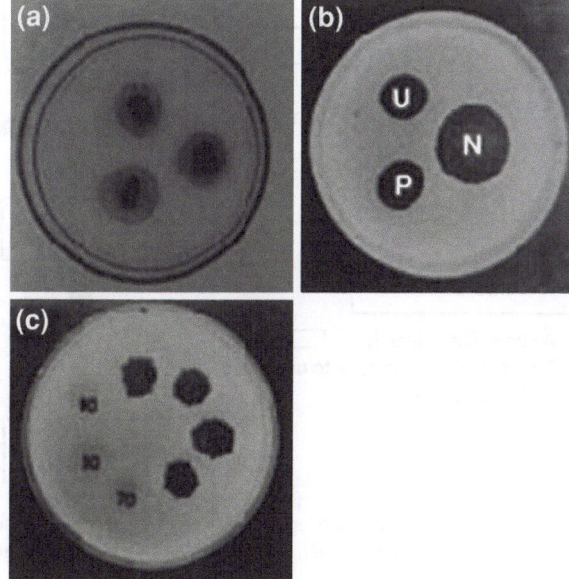

Fig. 7 Nattokinase three-dimensional (3D) structure and active site: Asp^{32}, His^{64}, Asn^{155}, and Ser^{221}

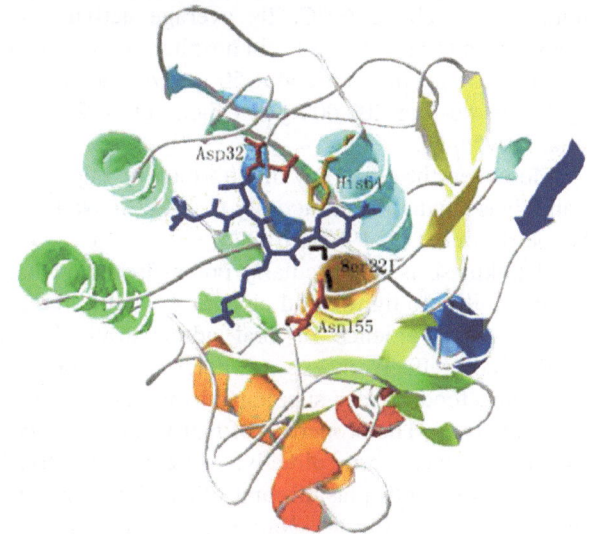

5.3 Structure and Functions

Nattokinase is a single-strand protein composed of 275 amino acids (Fig. 7) with molecular mass and *p*I of about 27,728 Da and 8.6, respectively. The enzyme has strong fibrinolytic activity at pH 6–12, but its fibrinolytic activity slows down for

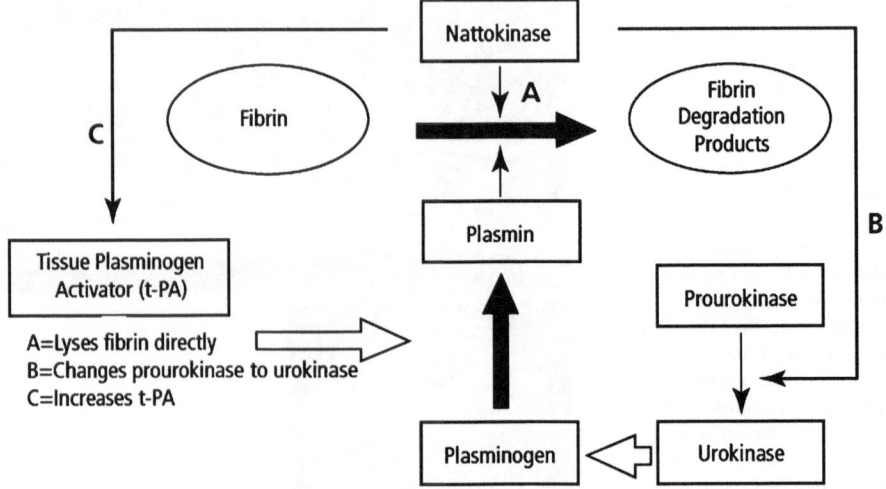

Fig. 8 Physiologic effects of nattokinase on fibrin. Nattokinase lyses fibrin directly (*A*), changes prourokinase to urokinase (*B*), and increases tissue plasminogen activator (t-PA)

temperatures above 60°C. Its average activity is calculated at about 40 CU (plasmin units)/g wet weight. In hospitals at present, urokinase is often utilized as a fibrinolytic enzyme, but about 50 g *natto* has the same efficacy as an amount of urokinase costing 200,000 yen (Zheng et al. 2005; Nakamura et al. 1992; Fujita et al. 1995a, b).

Moreover, the efficiency of a fibrinolytic injection lasts only 4–20 min, whereas nattokinase maintains its activity for 2–12 h. Nattokinase was easily extracted with saline (Fujita et al. 1995a, b).

Nattokinase is a particularly potent treatment because it enhances the body's natural ability to fight blood clots in several different ways and has many benefits including convenience of oral administration, confirmed efficacy, prolonged effects, cost effectiveness, and potential for preventative use. It is a naturally occurring, food dietary supplement that has demonstrated stability in the gastro-intestinal tract. The properties of nattokinase closely resemble those of plasmin in that it dissolves fibrin directly. More importantly, it also enhances the body's production of both plasmin and other clot-dissolving agents, including (endogenous) urokinase (Fig. 8). Nattokinase may actually be superior to conventional clot-dissolving drugs such as recombinant tissue plasminogen activators (t-PA), urokinase, and streptokinase, which are only effective therapeutically when taken intravenously within 12 h of a stroke or heart attack. Nattokinase, however, may help prevent the conditions leading to blood clots at an oral daily dose of as little as 2,000 fibrin units (FU) or 50 g *natto* (Sumi et al. 1987, 1990, 1992; Fujita et al. 1995a, b).

5.4 Prolonged Action

Nattokinase produces a prolonged action (unlike antithrombin drugs that wear off shortly after IV treatment is discontinued) in two ways: it prevents coagulation of blood, and it dissolves existing thrombus. Both the efficacy and the prolonged action of NK can be determined by measuring levels of euglobulin fibrinolytic activity (EFA) and fibrin degradation products (FDP), both of which become elevated as fibrin is dissolved. By measuring EFA and FDP levels, the activity of NK has been determined to last from 8 to 12 h. An additional parameter for confirming the action of NK following oral administration is a rise in blood levels of tissue plasminogen activator (t-PA) antigen, which indicates release of t-PA from endothelial cells and/or the liver (Kumada et al. 1994).

5.5 Thrombolytic Effect In Vivo

Nattokinase has been the subject of 17 studies, including two small human trials.

Fibrinolytic enzymes dissolve fibrin, the main component of blood clots, and medications containing these enzymes are the most effective methods used for treatment of thrombosis. There are two main screening tests of the clotting system: first, the prothrombin time (PT), i.e., the time taken by blood to clot, which is inversely proportional to the prothrombin concentration in blood; and second, the partial thromboplastin time (PTT) (occasionally the PTT is called the activated PTT or APTT). Both PT and PTT are clotting tests.

The fibrinolysis mechanism of NK has been explored more extensively than other microbial fibrinolytic enzymes. NK not only directly cleaves cross-linked fibrin, but also activates the production of t-PA, resulting in transformation of inactive plasminogen to active plasmin (Kumada et al. 1994; Fujita et al. 1995c).

Furthermore, NK enhances its fibrinolysis through cleavage and inactivation of plasminogen activator inhibitor-1 (PAI-1), which is the primary inhibitor of fibrinolysis and regulates total fibrinolytic activity by its relative ratio with t-PA (Urano et al. 2001). NK can be effectively absorbed across the rat intestinal tract after intraduodenal administration and then induces fibrinolysis (Fujita et al. 1995a). So far, NK is the only fibrinolytic enzyme from microorganisms whose thrombolytic effect in vivo has been best characterized.

Sumi et al. (1990) reported the effectiveness of *Bacillus subtilis* NK capsules in dissolving thrombi in dogs. After blood clots were experimentally induced in a major leg vein of male dogs, each dog was orally administered either four capsules (250 mg/capsule) of NK or placebo. They recorded that the blood clots in the dogs that received NK capsules completely dissolved within 5 h of treatment, and normal blood circulation was restored. However, as a negative control, blood clots in dogs that received the placebo did not show any sign of thrombolysis even after 18 h of treatment (Fig. 9).

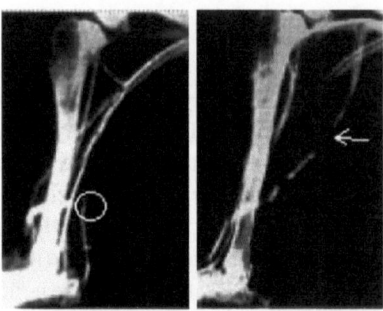

Fig. 9 X-rays of blood vessels of dogs that received nattokinase (*arrow*) or placebo. Dogs that received NK regained normal blood circulation free of the clot within 5 h of treatment. Blood clots in dogs that received only placebo showed no sign of dissolving in the 18 h following treatment

Moreover, dietary supplementation of *natto* can suppress intimal thickening and modulate lysis of mural thrombi after endothelial injury in rat femoral arteries (Suzuki et al. 2003b).

Lysis of the thrombus was observed by angiography (Kim et al. 1996a). More importantly, fibrinolytic activity, the amounts of t-PA, and FDP in the plasma are doubled when nattokinase is given to human subjects by oral administration.

The underlying mechanism involves the absorption of the administered *natto* enzyme across the intestinal tract, and the release of endogenous plasminogen activator that induces fibrinolysis in the occluded blood vessel (Sumi et al. 1990).

Fujita et al. (1995b) investigated the thrombolytic effect of NK on a thrombus in the common carotid of rat where the endothelial cells of the vessel wall had been injured by acetic acid. Animals treated with NK recovered 62% of arterial blood flow, whereas those treated with plasmin regained just 15.8%, and those treated with elastase did not exhibit any recovery. It was concluded that the in vivo thrombolytic activity of NK is stronger than that of plasmin or elastase.

Another human trial involved 12 healthy Japanese volunteers (6 men and 6 women, between 21 and 55 years old). Each participant had 200 g *natto* daily before breakfast, and their fibrinolytic activity was tracked through a series of blood plasma tests. The results recorded by Sumi et al. (1989) showed that oral administration of *natto*/nattokinase enhanced the ability of participants to dissolve blood clots. Furthermore, the volunteers retained such enhanced fibrinolytic activity for 2–8 h after administration. As a control, researchers later fed the same amount of boiled soybeans to the same volunteers and tracked their fibrinolytic activity. The tests showed no significant changes.

Lee et al. (2007) reported the antithrombotic effect of fibrinolytic enzymes on venous thrombosis to be due to the proteolytic enzymes absorbed into the blood stream prolonging the PT and APTT with the inhibition of aggregation of washed rat platelets induced by thrombin.

5.6 Benefits for Blood Pressure

Traditionally in Japan, *natto* has been consumed not only for cardiovascular support but also to lower blood pressure. In recent years, this traditional belief has been confirmed by several clinical trials. In 1995, researchers from Miyazaki Medical College and Kurashiki University of Science and Arts in Japan studied the effects of nattokinase on blood pressure in both animal and human subjects. In addition, the researchers confirmed the presence of inhibitors of angiotensin-converting enzyme (ACE), which converts angiotensin I to its active form angiotensin II. ACE causes blood vessels to narrow and blood pressure to rise; by inhibiting ACE, nattokinase has a lowering effect on blood pressure.

In an animal study, after a single IP of 400–450 g of 25 mg *natto* into rats, systolic blood pressure (SBP) significantly decreased from 166 to 145 mmHg in 2 h, and decreased further to 144 mmHg in 3 h.

The same *natto* extract was then tested on human volunteers with high blood pressure. Blood pressure levels were measured after 30 g lyophilized extract was administered orally for four consecutive days. SBP decreased on average from 173.8 to 154.8 mmHg. Diastolic blood pressure (DBP) decreased on average from 101.0 to 91.2 mmHg.

5.7 Preparation of Natto in Japan

There are several kinds of *natto* prepared in Japan, but here common *natto* is considered, which is a kind of vegetable cheese made by fermenting boiled soya beans, wrapped in straw and set in a warm cellar for 1 or 2 days (Muramatsu 1912). The product becomes white and mucilaginous through the development of bacteria. *Natto* is then consumed as a by-food after have been mixed with table salt and several stimulants such as powdered mustard.

There are several studies on *natto* constituents and the microorganisms growing on it, but no exact investigation about its preparation is known. So, manufacturers suffer from many difficulties in preparing *natto* of good quality. For this reason, Muramatsu (1912) was induced to make a study of the method of preparation. Besides, he thought it very useful to prepare *natto* of good quality and increase its consumption by people, as it is a very good and economical food stuff, being cheap and containing much protein, especially in a country where rice is the principal food.

Rice straw is used to wrap the boiled soya beans. Fresh straw is preferable to old straw, as its smell is better. The straw is cleaned by taking the muddy leaf away from the under part of the stem and then washing with clean water; afterwards, it is well packed at its two ends, leaving several inches apart, and bundled after filling the bag with the beans (Fig. 10). The reason for using straw for the preparation of *natto* is that it is considered to supply the proper bacteria to the beans. However, this is not considered the sole reason, as it can be prepared in other ways, for instance, by setting in a sterilized Petri dish or in a basket.

Fig. 10 *Natto* surrounded
with thick layers of straw

When it is made in a basket, which after filling with beans is set in a warm cellar covered with a straw mat, it is called basket *natto*. It is reasonable to consider the principal objectives of using straw for preparation of *natto* to be: (1) to supply of a good straw aroma to the *natto*, (2) to remove ammonia from the *natto*, and (3) to offer good air ventilation to the loosely packed beans.

The cellar for *natto* preparation is made from bricks or wooden piles, surrounded with thick layers of straw as walls, plastered with mud. The entrance is furnished with a thick door preventing air inlet.

Along the inside of the wall, a long shelf, 2 ft wide, is set up at a height of almost 2 ft, and one or two large hearths are made on the floor for the purpose of warming the room.

For the preparation of *natto*, the soya beans are first sorted, with all beans that are broken or imperfectly developed being picked out. After washing with clean water, they are soaked for several hours and boiled in an iron kettle until they become moderately soft (ca. 5 h). The boiled beans are enveloped with straw bundles while they are still hot, and the bundles are placed standing obliquely on the shelf in the cellar, which is previously warmed by charcoal to about 40°C. The cellar is then closed carefully to avoid air circulation. In this way, the beans become *natto* after 1 or 2 days and are ready for consumption.

The nutritional value of *natto* is greater than that of boiled soy beans, as it is rich in soluble and simple proteins and carbohydrates produced by microbial enzymes, especially proteolytic and sacchrolytic enzymes. So, when eaten with foods rich in protein or starch, they can be digested more rapidly than when eaten alone. In 1987, Sumi and his colleagues discovered NK in *natto*, which was then reported as the best fibrinolytic enzyme with thrombolytic activity.

5.8 Preparation of Natto in Microbiology Laboratories

Preparation of *natto* in the open air exposes the process to several contaminations and undesired effects due to mixed fermentation. So, it is necessary to perform this solid-state fermentation (SSF) under sterilized conditions.

Five hundred grams of soybeans is washed thoroughly, then soaked in 2 l tap water for 12 h. It is suggested that, when the water surface becomes bubbly due to germination, enough soaking has been done.

When the soybeans have absorbed enough water and swollen to twice their dry size, they are placed into a 2-l flask and steamed in a water bath for 2–5 h or in an autoclave at 121°C for 15 min. It must be ensured that the cooked soybeans are soft enough, otherwise the bacteria cannot penetrate to the center of the beans.

Soybean sauce (20 ml) is transferred into a sterile flask and mixed with 2 g natural salt, 4 ml molasses, and the growth content of the bacterial slant culture (e.g., *B. subtilis*).

The solution of salt, molasses, and bacteria is sprinkled over the soybeans, followed by stirring using sterile rubber gloves so that the bacteria are evenly distributed.

The result is transfered to wide Petri dishes, with sufficient humidity maintained in the atmosphere by filling two open plates with water, placing them on the middle incubator rack.

The temperature is maintained at 40°C for 24–48 h, until white "frost" is seen on the surface of the soybeans, and the laboratory is full of the aroma of *natto*. Some smell of ammonia is normal, but if it is too strong, undesirable germs may have flourished.

At the end of incubation, the *natto* is kept in the refrigerator for a few days to 1 week for aging.

The *natto* is washed with water before eating to remove the viscous layer, where bacteria are present. The bacterium synthesizes also vitamin K in the form of menaquinone (MK). MK serves as an essential cofactor of vitamin K-dependent carboxylase, which catalyzes synthesis of clotting factors II, VII, IX, and X and protein C and S, which may interfere with the anticlotting action of NK (Suttie and Machlin 1991).

5.9 Contraindications and Precautions

5.9.1 Thrombic Embolization

Because *natto* is a very potent fibrinolytic agent, it is theoretically possible that regular use could break a clot loose from a lower-extremity vein, which could cause a pulmonary embolism or pass upward to the brain and cause a stroke. Although no controlled studies have been conducted to demonstrate *natto*'s safety with regard to avoiding clot mobilization, historically, documentation records that *natto* is significantly safe.

5.9.2 Interaction with Coumadin

The nattokinase in *natto* could require a decrease of patients' doses of coumadin due to drug interaction, whether a patient eats *natto* or takes nattokinase extract. While on coumadin, it is necessary to take a consistent amount of *natto*/nattokinase each day. Physicians also need to monitor clotting time [PT, PTT, and international normalized ratio (INR)] in the first weeks of *natto* or nattokinase therapy until these levels are stable.

5.9.3 Intervention with Warfarin

The critical role of vitamin K for blood clotting is well recognized (Suttie and Machlin 1991). Vitamin K serves as an essential cofactor of vitamin K-dependent carboxylase, which catalyzes synthesis of clotting factors II, VII, IX, and X and protein C and S (Maillard et al. 1992). In addition, patients who are taking warfarin for prevention of thromboembolism must restrict their intake of vitamin K, since the anticoagulant action of warfarin depends on inhibition of the vitamin K-dependent carboxylation reaction. *Natto* is strongly prohibited for those patients, because it contains high levels of vitamin K. *Natto* also contains living *Bacillus subtilis*, which continues to synthesize vitamin K in the intestine for several days after *natto* has been eaten in the form of menaquinones (MKs) (Kaneki et al. 2001).

So, it is important to find a suitable cooking method for *natto* to decrease the content of bacilli, so that patients on warfarin can eat *natto* without encountering excessive plasma concentrations of vitamin K.

Homma et al. (2006) developed five cooking methods to decrease the number of *Bacillus subtilis* in *natto*. Thirty-two healthy volunteers (19 males and 13 females) aged from 21 to 60 years participated in the study. Plasma concentrations of vitamin K were determined before and after the subjects had ingested *natto*. The numbers of *Bacillus subtilis* in feces were also examined. Cooking methods were (1) washing: *natto* was washed with water at room temperature to remove the viscous layer from soybeans, since *Bacillus subtilis* resides on the surface; (2) single heat treatment: *natto* was treated with steam at 100°C for 15 min; (3) double heat treatment: *natto* was treated with steam at 100°C twice for 15 min each with a 60-min interval; (4) double boiling with short interval: *natto* was boiled at 100°C for 15 min twice with 60-min interval; and (5) double boiling with a long interval: *natto* was boiled twice at 100°C for 15 min each with a 12-h interval. In conclusion, boiled *natto* did not cause a marked increase in the plasma concentration of vitamin K in subjects who consumed it. Thus, patients on warfarin may be able to eat boiled *natto* without undesired effects.

Conclusions

The traditional Japanese food *natto* has been used safely for over 1,000 years. The potent fibrinolytic enzyme nattokinase appears to be safe based upon the long-term traditional use of this food. Nattokinase has many benefits including convenience of oral administration, confirmed efficacy, prolonged effects, cost effectiveness, and potential for preventative use. It is a naturally occurring food-based dietary supplement being marketed in this country by a US nutraceutical firm, and has demonstrated stability in the gastrointestinal tract as well as to changes in pH and temperature.

All prior clinical research points to nattokinase's effectiveness and safety for managing a wide range of diseases, including hypertension, atherosclerosis, coronary artery disease, stroke, and peripheral vascular disease. Evidence from long-term use at high doses in Japanese people points to nattokinase as a safe nutrient that acts as a very powerful fibrinolytic agent. However, more research on humans is needed to verify the predicted safety of formulated extracts that deliver high concentrations of nattokinase while eliminating naturally occurring vitamin K.

6 Assessment of Fibrinnolytic Activity

Since 1954, when the activity of plasmin and thrombin on synthetic substrates were first demonstrated (Sherry and Troll 1954; Troll et al. 1954), synthetic substrates have played an important role in the study and characterization of various enzymes involved in blood fibrinolysis (Abloundi and Hagan 1957; Kline and Fishman 1961).

Activation of plasminogen to plasmin involves an enzymatic proteolytic step, and the activators mediating the action share the ability of plasmin to split arginine and lysine esters (Alkjaersig et al. 1958).

Existing assay methods can be divided into three major groups:

1. The first group consists of indirect assays of fibrinolytic activity with protein as substrates, e.g., the fibrin plate method (Haverkate and Bradman 1975), clot lysis method (Lassen 1958), and caseinolytic methods (Kline 1971). PAs cannot be assayed directly by these methods but only via their activating action on plasminogen present in or added to the substrate. With the fibrin plate method extremely low levels of enzymatic activity can be determined, but long incubation times are required, and the responses found depend on the quality of the fibrin substrate (Haverkate and Bradman 1975).
2. A second group of assays, suitable for kinetic studies, is based on the determination of hydrolysis of synthetic peptide esters (Bell et al. 1974). The disadvantage of the use of these esters is that esterolytic rather than amidolytic activity of a proteolytic enzyme is measured.
3. In the third group of synthetic substrates, e.g., acetyl L-lysine *p*-nitroanilines, amidolysis of the peptide amides is measured (Petkov et al. 1973).

6.1 Indirect Assays

The standard system used for measuring PA in cells is an indirect, two-step assay in which plasminogen is incubated with a source of PA and the plasmin activity generated is quantitated by using fibrin, casein, or protamine as substrate (Goldberg 1974; Unkeless et al. 1974; Kessner and Troll 1976).

Marsh and Gaffney (1977) developed a rapid fibrin plate method for plasminogen activator assay. Their study was carried out to investigate plasminogen enrichment as a means of shortening the incubation period, which is associated with the fibrin plate method. Fibrin plates were made up to contain two casein units of added plasminogen. Each was opaque, firm, did not lyse spontaneously, and yielded biometrically valid parallel-line assays for SK and UK.

Jespersen and Astrup (1983) described the reproducibility, precision, and required conditions of the fibrin plate method for determination of fibrinolytic agents. Under optimal conditions the assay is a sensitive and precise method for quantitative determination of fibrinolytic agents.

Millar and Smith (1983) compared the rapid and highly sensitive solid-phase assay with the fibrin plate method for measurement of UK, SK, and PA in human euglobulin fractions. The solid-phase assay was run using Glu- or Lys-plasminogen, and significant differences were observed in the activation of plasminogen by UK and SK. Very good agreement was obtained between the fibrin plate and solid-phase methods in all cases.

Fossum and Hoem (1996) developed a fibrin microplate method for estimation of UK and non-UK fibrinolytic activity in protease-inhibitor-deprived plasma. In this method, fibrin clots, with a suitable dye incorporated, were formed in wells of standard high-adsorption microtiter plates.

Roche et al. (1983) presented a rapid and highly sensitive solid-phase radio assay for measurement of PAs. The method employs a convenient and stable ^{125}I-fibrinogen latex-bead product and can reproducibly detect 0.25 milli PU/ml of fibrinolytic enzyme. This represents a 100-fold increase in sensitivity of UK over the radioisotopic solid-phase technique and a 120-fold increase over the sensitivity of the fibrin plate method.

6.1.1 Fibrin Plate Method

The fibrin plate (Astrup and Mullertz 1952) has two forms: the plasminogen-free fibrin plate and the plasminogen-rich fibrin plate.

The plasminogen-free fibrin plate is made up of fibrinogen solution [5 mg human fibrinogen in 7 ml 0.1 M barbital buffer (pH 7.8), 10 U thrombin solution, and 7 ml of 10 g agarose L^{-1}] in a Petri dish (9 cm diameter). This Petri dish is heated at 80°C for 30 min to destroy other fibrinolytic factors. Compared with the plasminogen-free fibrin plate method, the plasminogen-rich fibrin method also includes 5 U plasminogen and is not heated.

The heated fibrin plate (Lassen 1953) was prepared using bovine fibrinogen and thrombin, again modified by inclusion of calcium or sodium chloride in the plate.

To observe the fibrinolytic activity of the enzymes, 10 μL enzyme solution is carefully dropped on the fibrin plate and incubated at 37°C for 18 h. The activity of a fibrinolytic enzyme was determined by measuring the diameter of the clear zone on the fibrin plate and plotting a standard curve calculated using different concentrations of urokinase as a standard fibrinolytic enzyme.

6.1.2 Dyed Fibrin Plate Assay

Barta (1966) developed a fibrin plate test for assaying fibrinolytic activity by dyeing the fibrin with Congo Red and applying the protease solution to the entire surface of the fibrin film. Enzymatic digestion results in uptake of the dye by the supernatant solution, which is then measured by colorimetry. The applicability of the test for rapid and accurate evaluation of protease activity is demonstrated by the dose–response curves of enzymatic digestion, and their standard errors.

6.1.3 Plasma Streptokinase Lysis Time (PSLT)

In this method, 100 μl streptokinase at concentration of 100, 150, 200, 500, 1,000, or 2,500 U/ml of plasma is added to test-tubes containing 100 μl plasma and 800 μl phosphate buffer, and the mixture is clotted by addition of 5 U stock solution of thrombin. The lysis time is defined as the time elapsed from the moment of adding thrombin until complete lysis of the clot (Gidron et al. 1978).

6.1.4 Streptokinase-Activated Lysis Time (SALT)

In this method, blood samples are centrifuged at room temperature for 30 min at 3,000 rpm, and the plasma separated. Then, 0.15 ml of a 100 U/ml solution of streptokinase is added to test-tubes containing 0.1 ml plasma and 0.65 ml phosphate buffer, resulting in a final concentration of 150 U SK per ml of plasma. The mixture is clotted by addition of 0.1 ml stock solution of thrombin. The end-point of lysis is determined visually or by a nephelometric method (Gidron et al. 1978). The lysis time is defined as for the plasma streptokinase time above.

6.1.5 Nephelometric Method for Recording Lysis Time

A fluorimeter with a recorder is used to register the clotting and lysis of plasma by SALT method. A 400 nm excitation filter with a 1% neutral density filter is used without an emission filter. The different light scattering properties of the plasma solution and the clot permit the plotting of coagulation and lysis. The fluorimeter is calibrated using the plasma-SK mixture, and the return of the curve to baseline is used as the end point (Gidron et al. 1978). There is close agreement between the visual and nephelometric SALT methods.

6.1.6 Dilute Blood Clot Lysis Time (DBCLT)

This method is essentially that of Chakrabarti and Fearnley (1962). For each of the three diluents investigated (phosphate buffer 2, 0.12 M sodium acetate, and 0.12 M sodium chloride) a series of seven blood dilutions is prepared, arranged to give final blood concentrations of 5, 7.5, 10.0, 12.5, 15, 17.5, and 20%. The total volume of each test system is 2.0 ml, including 0.1 ml thrombin solution (50 units/ml) diluted in the appropriate salt solution. After mixing the contents, the tubes are transferred to a 37°C water bath. Approximately 10 min after the addition of thrombin, the tubes are rotated briskly between the palms of the hands to release clots from the sides of the tubes. Lysis times are recorded when the clots have disintegrated completely.

Results of such experiments by Gallimore (1967) indicated that sodium acetate at 0.12 M and pH 7.4 is to be preferred to phosphate buffer at the same pH and ionic strength as a diluent for the dilute blood clot assay. Phosphate ions potentiate inhibition more than acetate ions and lead to longer lysis times. The results also indicate that diluted blood and plasma clots lyse, not only because of the dissociation of a plasmin–antiplasmin complex as postulated by Macfarlane and Piling (1946), but also by virtue of the dilution of chloride ions in the blood or plasma.

6.1.7 Euglobulin Clot Lysis Time (ECLT)

The euglobulin clot lysis time is a measure of plasminogen activator activity which is performed at −4°C (Walker and Davidson 1985). Plasma is collected and anticoagulated with 0.105 M sodium citrate, stored at −70°C, and thawed to 4°C immediately before use. Separation of the euglobulin fraction, containing fibrinogen, plasminogen, and plasminogen activator, was achieved by dilution of 0.5 ml chilled anticoagulated plasma in 9 ml distilled water and acidification with 0.1 ml 1% acetic acid. After standing at 4°C and centrifugation at 3,000 rpm for 10 min, the euglobulin fraction (certain clotting factors) precipitates, the supernatant is discarded, and the precipitate is resuspended in borate buffer (pH 9). Aliquots (0.2 ml) of the suspension are clotted with 0.2 ml 0.025 m calcium chloride or 0.1 ml thrombin. The time taken for complete dissolution of the clot at 37°C is called the euglobulin lysis time.

Howell (1964) introduced a simple method for measuring partial clot lysis using cheap and easily available apparatus and reagents. The method was first described by Blix (1962) and has been adapted to suit the apparatus available and the subject being investigated.

The principle is as follows: a fibrin clot containing red blood cells is made and exposed to a fibrinolytic enzyme. As the clot lyses, red cells are released from it, and the amount of lysis occurring in a given time can be found by measuring the number of cells released.

A clot of 1 ml volume is made in a lusteroid tube of 11×54 mm. The basic components are 0.1 ml fibrinogen (1% solution) contaminated with plasminogen, 0.1 ml thrombin (50 units/ml), 0.1 ml red blood cells (platelet-free, having been washed several times), and 0.7 ml buffer space. The clot is buffered to pH 7.2. Group O red cells are used, because human serum fractions added to the clot may contain blood group antibodies. The 0.7 ml buffer space is used for addition of components of the fibrinolytic system so that their effect on clot lysis can be measured. Also, any substance whose function in the fibrinolytic system is unknown can be substituted for the buffer space for assessment of its activity as an inhibitor or an enzyme.

To ensure that the red cells are distributed evenly throughout the clot, fibrinogen is added last, and the tube is stoppered and inverted rapidly three times. The tube is then left standing upside down on its stopper, thus making it easy to extract the clot without breaking it. After at least 30 min, the clot is gently transferred to a Petri dish containing physiological saline in order to remove any loose red cells. From the Petri dish, it is moved to a polythene container measuring 13×30 mm with a push-on lid and containing 2 ml of fluid in which the clot floats. The nature of this fluid depends upon the experiment being carried out. It may be buffer, any known component of the fibrinolytic system, or any substance whose effect on fibrinolysis one wishes to discover.

The polythene container and clot are now placed in a 37°C incubator on a rotating device. Blix (1962) used a record player, but a Matburn mixer is quite suitable. As the clot lyses, red cells are released into the surrounding fluid, and after a certain time, the clot is lifted out of the container and the number of red cells in the surrounding fluid measured. To do this 0.5 ml of the fluid containing the red cells is transferred to 4 ml 0.04% ammonium hydroxide and read in a spectrophotometer at wavelength of 542 nm against a blank in which 0.5 ml saline is used in place of the red cell suspension. The reading obtained on the spectrophotometer is converted to the percentage of total clot lysis by calculating a calibration curve using the same batch of red cells as was used in the clot, in which the 100% value is equivalent to all the red cells in the clot being released into the surrounding fluid.

Hawkey and Stafford (1964) developed a standard clot method for assay of plasminogen activators, antiactivators, and plasmin, which is shorter and easier to use quantitatively than the fibrin plate method (Astrup and Mullertz 1952), which has the further disadvantage that the test substance cannot actually be incorporated into the clot. Neither the special apparatus required for the ^{131}I-labeled clot procedure nor the preparations involved in the standard thrombus (Blix 1962) are necessary.

Basically, the clot consists of 0.2 ml fibrinogen and 0.1 ml thrombin, each in a prestandardized solution, leaving a reagent space of 0.8 ml into which plasminogen, activator, and test enzymes or substances can be introduced as desired, the volume being made up by saline phosphate buffer.

The test is carried out in 75 × 12 mm glass test-tubes which have been washed in detergent, rinsed, soaked overnight in 2% HCI, and rerinsed thoroughly with tap and then distilled water. Tubes and reagents are placed in an ice bath. Reagents are added as rapidly as possible in the sequence saline buffer, plasminogen, fibrinogen, activator, and finally thrombin; several tubes can be set up together, providing that the total length of time spent in preparation is not more than 1.5 min. A stopwatch is started as the thrombin is added, the contents of the tube are quickly mixed, preferably by means of an automatic mixer, and the tubes transferred to a water bath at 37°C. One minute later, they are examined to confirm that clotting has taken place; thereafter they are examined every 30 s, or at regular intervals determined by experience of the particular test being carried out. The end-point is taken as the time of complete disappearance of the clot and is recorded as the lysis time in minutes.

Somerville (1972) performed clot lysis in 50 × 6 mm test-tubes. In each sterile test-tube were placed 0.5 ml fibrinogen (1% in veronal buffer, pH 7.4) and 0.02 ml 0.1 M calcium chloride. All solutions were sterilized by filtration before use. Finally, 0.02 ml sterile bovine thrombin (0.5 units) was added. Thorough mixing was ensured using a vortex stirrer, and the resulting fibrin clot in which the test strain filtrate was suspended was incubated at 37°C and examined after 4.5 , 20 h, and then daily until control tubes showed evidence of clot disintegration. Bovine and human fibrinogen were used, and the control fibrin clots were stable for more than 4 days in the case of the bovine fibrinogen and over 3 days for the human fibrinogen. Culture filtrates of a limited number of strains were also tested, and the results obtained with these were similar to those obtained with the organisms themselves.

6.1.8 Effect of Calcium and Decalcifying Agents on Assays

The inhibitory effect of calcium on lysis in plasma (Sherry et al. 1959), and its accelerating effect on euglobulin clot lysis (Ratnoff 1952), have been confirmed.

It is well known that metallic ions acting as cofactors can play an important role in enzymatic reactions. In the absence of inhibitors, lysis of fibrin proceeds more rapidly when calcium is present. It has not been determined whether calcium accelerates the proteolytic activity of plasmin itself, or influences the activation of plasminogen to plasmin by activator. Ratnoff (1952) believed that activator is affected directly. The elucidation of this problem is made difficult by the fact that any test ultimately depends on the proteolytic activity of plasmin, and thus it is hard to distinguish between activator and plasminogen on the one hand and plasmin on the other.

In the light of the results using the euglobulin fraction, it was thought that the inhibitory effect of calcium in clot lysis could not be due merely to the different type of fibrin formed in the presence of calcium as suggested by Bickford and Sokolow (1961). They pointed out that the sulfhydryl-bonded fibrin formed in the presence of calcium is insoluble in urea, and put forward the hypothesis that this sulfhydryl-bonded fibrin might also be more resistant to fibrinolytic enzymes. This difference in solubility of fibrin in urea is dependent on both calcium and the fibrin-stabilizing factor (Laki and Lorand 1948), and it may be that this factor plays some part in making the clot more resistant to the fibrinolytic enzymes.

Calcium may inhibit fibrinolysis by acting through inhibitors present in whole blood or plasma.

According to Flute (1960), antiactivator is formed when blood is incubated in contact with glass, and is precipitated with the euglobulin fraction. The inhibition it caused was not enhanced by calcium. However, as reported by Flute PT (1962, personal communication), it was found that the formation in vitro of antiactivator requires calcium. This would explain the importance of the rapid dilution of whole blood at 4°C in phosphate buffer in the 1 in 10 clot lysis time, so that calcium is removed before antiactivator can be formed (Fearnley et al. 1957).

The antiplasmin activity of plasma was found to be enhanced by calcium, and this would contribute to the inhibition by calcium of fibrinolysis in whole blood or plasma. It is not known how calcium influences these inhibitors, or whether its effect is fully explained by this mechanism.

The finding that calcium prevents spontaneous lysis of unheated fibrin plates prepared from some batches of Armour bovine plasma fraction I is difficult to explain in the light of these results. Fraction I is contaminated by plasminogen and traces of activator; antiplasmins are also present, but it is doubtful whether at sufficient quantity to account fully for the inhibition of lysis when calcium is added. In such instances it is possible that the effect of calcium on the type of fibrin formed is of predominant importance, due to the lower concentration of the enzymes and their inhibitors.

It is apparent that the mechanism of inhibition of fibrinolysis by calcium in vitro is complex. It is partly explained by the effect of calcium on the formation of antiactivator and on antiplasmin activity, and is possibly influenced to some extent by the type of fibrin formed when calcium is present. In whole blood, these inhibitory influences must modify the accelerating effect of calcium observed when inhibitor-free euglobulin fraction is used.

Decalcifying agents also may affect clotting assays and are (1) 10% EDTA (0.1/10 ml blood), (2) 3.8% trisodium citrate (1/10 ml blood), and (3) 0.1 M ammonium oxalate (1/10 ml blood). The EDTA interfered to some extent with the subsequent clotting procedures, and, as citrate was found to be suitable for removal of calcium in these experiments, this was used in the majority of the tests.

Buckell (1958) found that citrate may be active in the transformation of proactivator to activator, which in turn activates plasminogen to plasmin, with this increased plasmin content being shown on both the unheated and the heated plates and by the increased speed of lysis. Further work needs to be done to determine the precise site of action of citrate, but it appears that citrate should be avoided in the preparation of specimens for estimation of fibrinolysis by either the euglobulin lysis time method or the fibrin plate method, and that oxalate is suitable as it gives a result very close to that of true plasma.

Buckell (1958) also found oxalate to be a suitable anticoagulant for use in the above methods.

6.2 Esterolytic Assays

Sherry et al. (1966) investigated the ability of UK as a fibrinolytic enzyme to hydrolyze a variety of α-amino substituted Arg and Lys esters (acetyl–Arg methyl ester, benzoyl–Arg methyl ester, tosyl–Arg methyl ester, Lys methyl ester, acetyl–Lys methyl ester, benzoyl–Lys methyl ester, and tosyl–Lys methyl ester). Their observations indicated that UK catalyzes more rapid hydrolysis of lysine esters and its derivatives than the corresponding esters of arginine. Substitution of α-amino group of lysine methyl esters increases the sensitivity of the ester to hydrolysis. They further reported that acetyl–Lys methyl ester is the most sensitive substrate among the various esters tested.

A convenient and highly sensitive colorimetric assay for various proteases such as trypsin, chymotrypsin, plasmin, thrombin, and UK has been reported (Ninobe et al. 1980). The substrates used were naphthyl ester derivatives of N-tosyl L-lysine, N-acetyl glycyl L-lysine, and N-acetyl L-tyrosine. Activity was assayed by colorimetric determination of naphthol released. It was reported that this method is more sensitive than use of corresponding methyl or ethyl ester derivatives.

Barlow and Marder (1980) reported the use of a chromogenic substrate L-pyroglutamyl glycyl L-arginine p-nitroanilide (S-2444) for assay of plasma UK levels of patients treated with UK of urinary or tissue culture source. The p-nitroaniline released was measured by spectrophotometer at 405 nm. A linear response relationship between UK concentration and optical activity was obtained, indicating that the method detects UK in quantitative manner.

Kulseth and Helgeland (1993) developed a simple and highly sensitive chromogenic microplate assay for quantification of rat and human plasminogen in plasma samples and subcellular fractions. The assay is based on conversion of plasminogen to plasmin using UK or the tested enzyme as an activator, and subsequent cleavage of the chromogenic plasmin substrate D-alanyl-L-cyclohexylanyl-L-lysine-p-nitroanilide dihydroacetate.

p-Nitroaniline released by the cleavage is then measured at 410 nm by microplate reader. The assay includes an acidification step to make plasminogen more readily activated to plasmin. The method is suitable for analyses of a large number of samples, measuring plasminogen in the nanogram range (0.5–50 ng/50 μl of sample).

6.3 Fluorimetric Assays

Kessner and Troll (1976) reported a new method for determining plasminogen activator levels. The assay is based on digestion of N-terminal blocked protamine and subsequent measurement of the exposed amino groups using the fluorogenic amine reagent, fluram.

Nieuwenhuizen et al. (1978) reported fluorogenic substrates for sensitive and differential estimation of UK and t-PA. Two fluorogenic peptide amides have been synthesized, i.e., Boc L-valyl-glycyl-L-arginine-L-naphthylamide and L-valyl-glycyl-L-arginine-2-naphthylamide. The kinetic parameters of plasmin, UK, and human uterine tissue plasminogen activator on substrates 1 and 11 have been determined.

Zimmerman et al. (1978) developed a simple, sensitive, direct assay that allowed both rapid measurement and kinetic analysis of PA, independent of plasmin generation. The method employed a synthetic fluorogenic peptide substrate 7-(N-CbZ-glycylglycyl argininamido)-4-methylcoumarin trifluoroacetate. The assay correlated well with the standard ^{125}I-labeled fibrin plate assay using highly purified UK.

7 Hemostasis Screening Tests

Laboratories usually perform a set of tests that aims to identify most clinically important clotting defects. Invariably this includes PT, APTT, fibrinogen, and usually thrombin time. It is important to perform a full blood count to quantify the platelets. When combined with results of the platelet count obtained from the complete blood count (CBC), differential diagnoses can be generated to assist in evaluation of patients with clotting or bleeding disorders.

7.1 CBC Assay

Typically, EDTA-anticoagulated blood is obtained for analysis in an automated particle counter. The reported platelet count is usually quite precise [coefficient of variance (CV) ~5%]. In asymptomatic patients in whom thrombocytopenia is reported, the possibility of pseudothrombocytopenia or EDTA-induced thrombocytopenia should be considered, especially in patients without history of bleeding. This phenomenon occurs in 0.1–1% of normal people; it results from EDTA modifying platelet membrane proteins which then react with preexisting antibodies present in patient blood that recognize the modified platelet proteins, producing platelet clumping or satellitism (Fig. 11). It should be routine laboratory policy for technical personnel to review peripheral blood smears of patients with newly diagnosed thrombocytopenia to determine whether the thrombocytopenia is true or spurious. If EDTA-induced thrombocytopenia is suspected, the CBC should be repeated using blood collected in a citrate or acid-citrate-dextrose collection tube. Figure 12 illustrates an algorithm that suggests one strategy to evaluate thrombocytopenia. In terms of hemostasis evaluation, one limitation of the CBC is that,

Fig. 11 Platelet satellitism and platelet clumping in a blood smear from a patient with EDTA pseudothrombocytopenia

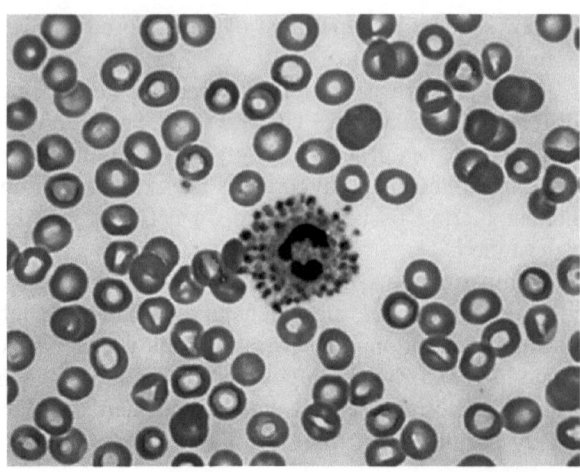

even though it is usually a reliable indicator of platelet count, it does not measure platelet function. The bleeding time test was originally thought to perform this function, but as discussed below, it is not uniformly reliable in assessing platelet function.

7.2 PT Assay

The PT assay has two purposes: to screen for inherited or acquired deficiencies in the extrinsic pathway of coagulation (Fig. 13) and to monitor oral anticoagulant therapy. The PT is affected by decreased levels of fibrinogen, prothrombin, or factors V, VII, or X. Since three of the five coagulation factors measured by the PT are vitamin K-dependent proteins (prothrombin, and factors VII and X), the PT assay is useful in detecting vitamin K deficiency from any cause, including liver disease, malnutrition, or warfarin therapy. PT does not measure factor XIII activity or components of the intrinsic pathway (Rodgers 2004). The PT assay is performed by mixing patient plasma with thromboplastin. Thromboplastin is a commercial tissue factor–phospholipid–calcium preparation, which is derived either from animal tissue or from recombinant methods. Tissue factor in the thromboplastin preparation binds factor VII in patient plasma to initiate coagulation. The clotting time is measured in seconds using instruments with mechanical or photooptical endpoints that detect fibrin formation (Hougie 1990). Thromboplastin preparations can vary in their sensitivities, resulting in different clotting times. A typical PT reference range is 10–15 s. Most PT assays are automated, with the instrument adding reagents and patient plasma samples per protocol. In general, the PT assay is more sensitive in detecting low levels of factors VII and X than low levels of fibrinogen, prothrombin, or factor V. In particular, different thromboplastin

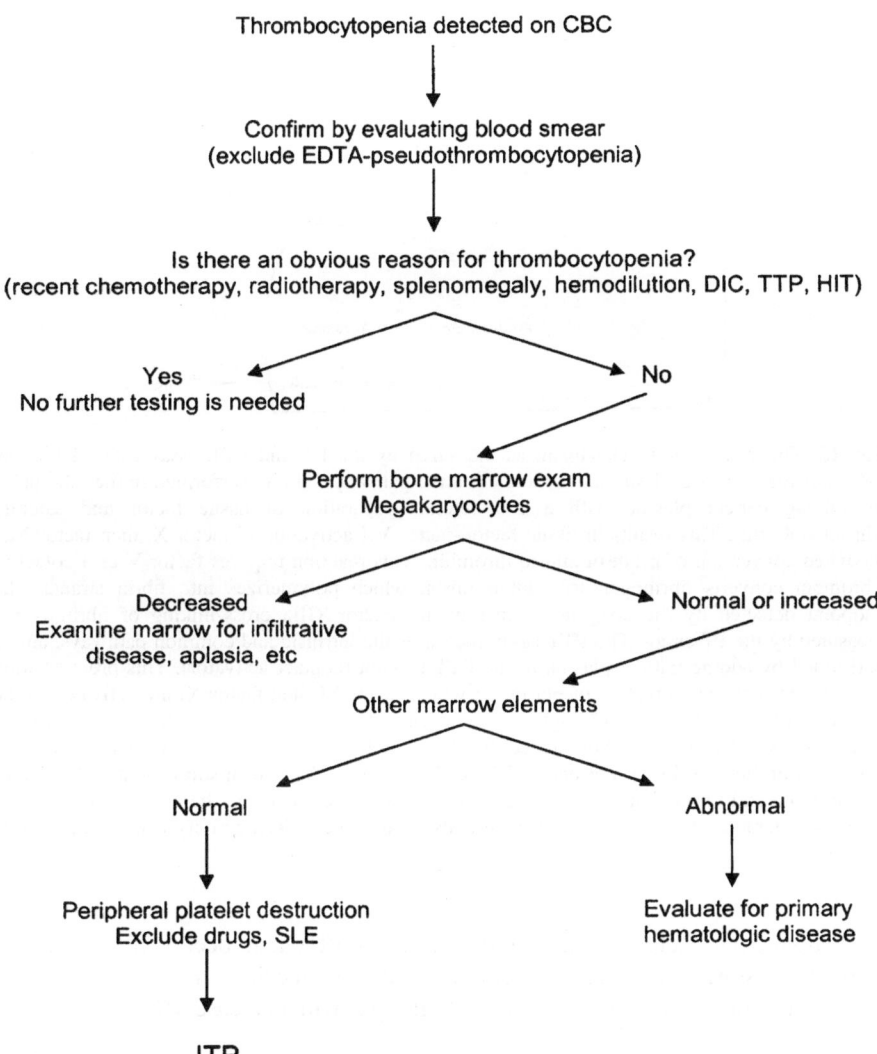

Fig. 12   A strategy to evaluate thrombocytopenia

reagents may exhibit variable sensitivities to these factor deficiencies. Mild factor deficiency (i.e., 40–50% of normal) may not be detected by many thromboplastin reagents. The PT assay is less affected by heparin than is the PTT assay, but therapeutic doses of unfractionated heparin will usually prolong PT by a few seconds unless a heparin neutralizer is present in the PT reagent (Hougie 1990; Rodgers 2004). Shortened PT values may result from either poor-quality venipuncture (activated sample), chronic disseminated intravascular coagulation (in vivo activation), or cold activation of the sample (in vitro

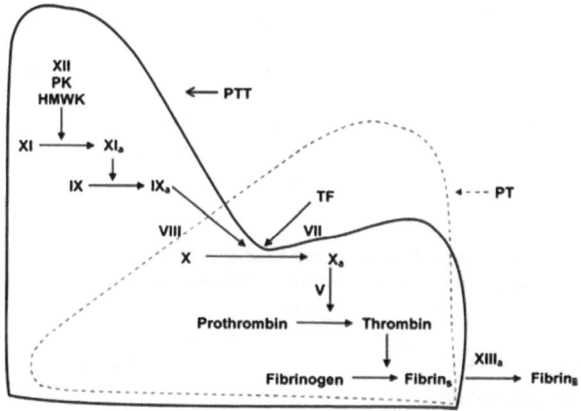

Fig. 13 The coagulation mechanism as measured by the PT and PTT assays. The PT assay measures the extrinsic (tissue factor) and common pathways, and is performed in the laboratory by mixing patient plasma with a commercial preparation of tissue factor and calcium (thromboplastin). This results in tissue factor–factor VII activation of factor X, then factor Xa–mediated conversion of prothrombin to thrombin. This reaction requires factor V as a cofactor. Thrombin converts fibrinogen to soluble fibrin, which polymerizes into fibrin strands, the endpoint detected by the coagulation instrument. Factor XIIIa cross-linking of fibrin is not measured by the PT assay. The PTT assay measures the intrinsic and common pathways, and is performed by adding patient plasma to the PTT reagent (contact activator). This preincubation step initiates contact activation of plasma in which factor XII and factor XI are activated in the presence of prekallikrein and high-molecular-weight kininogen. Factor XIa then activates factor IX–IXa. Calcium is then added to the sample. This results in factor IXa-mediated activation of factor X in the presence of the cofactor, factor VIII, with subsequent activation of prothrombin and fibrinogen as described above. As with the PT assay, the endpoint of the PTT assay is generation of polymerized fibrin strands, so that factor XIII activity is not measured by the PTT assay

activation from factor XII activation of factor VII) that occurs if the plasma sample is stored at cold temperatures (above freezing) for several hours. Administration of recombinant factor VIIa also will decrease PT.

7.3 PTT Assay

The partial thromboplastin time assay is useful for three reasons: as a screening test for inherited or acquired deficiencies of the intrinsic pathway (Fig. 13), to detect the lupus anticoagulant, and to monitor heparin therapy. PTT is affected by decreased levels of intrinsic pathway components (factors VIII, IX, XI, and XII, prekallikrein, and high-molecular-weight kininogen), as well as decreased levels of common pathway components (fibrinogen, prothrombin, and factors V and X). Factors VII and XIII are not measured by the PTT assay (Rodgers 2004). To perform the PTT assay, patient plasma is preincubated with the PTT reagent (crude phospholipid and a surface-activating agent such as silica or kaolin). This preincubation initiates contact activation (intrinsic pathway activation) in which factors XII and XI are activated in the presence of cofactors, prekallikrein, and high-molecular-weight kininogen. Factor XIa then converts factor IX to IXa. Calcium is then added to the preincubation mixture; this results in factors IXa/VIII activation of factor X, then factor Xa/V-mediated activation of prothrombin to thrombin followed by conversion of fibrinogen to soluble fibrin that polymerizes into fibrin strands, the endpoint of the PTT assay (Hougie 1990; Rodgers 2004). As with the PT assay, most PTT assays are automated. Mild factor deficiency (i.e., 30–50% of normal) may not be detected by most PTT reagents, but deficient levels of 10–20% should be detected. The usual PTT reagent is less sensitive to factor IX than to factors VIII, XI, and XII. The ability to detect mild factor deficiency is reagent dependent, and this is an important consideration when choosing the laboratory PTT reagent; for example, a hospital laboratory that evaluates a large population of bleeding disorder patients may prefer a PTT reagent that is more sensitive to detection of factor deficiency than to detection of lupus anticoagulants. In contrast, a hospital laboratory that evaluates large numbers of obstetrical patients may prefer a PTT reagent that has the opposite characteristics. PTT may be affected by high levels of factor VIII, an acute-phase response protein; high factor VIII levels may mask coexisting mild intrinsic coagulation deficiencies. A typical PTT reference range is 25–36 s. Shortened PTT values may result from poor-quality venipuncture (activated sample), chronic disseminated intravascular coagulation (in vivo activation), or increased factor VIII levels. It should be emphasized that the PT and PTT assays are screening tests, that normal PT/PTT results do not exclude a bleeding disorder, and that many patients with mild factor deficiency will have normal results for these assays (Rodgers 2004). In the mixing test (inhibitor screen), abnormal PT or PTT results are due either to deficiency of a factor measured by the assay, or by an inhibitor such as an antibody or heparin. Uncommon inhibitors include fibrin degradation products and certain paraproteins. The mixing test is useful to distinguish between deficiency versus inhibitor, and mixing test results usually suggest subsequent test ordering. The most common mixing test protocol mixes patient plasma with normal plasma in a 1:1 ratio, followed by repeat assay of the PT or PTT immediately after mixing and repeated 1–2 h later. Sample mixes that correct to normal at both intervals suggest that the original abnormal result was due to factor deficiency, while sample mixes that fail to correct to normal at one or both intervals suggest the presence of an inhibitory substance (Rodgers 2004). If heparin is suspected as the inhibitor, screening tests for the presence of heparin can be done with the thrombin time assay and reptilase assay, or by direct assay of heparin or low-molecular-weight heparin using anti-factor Xa methods. The mixing test is most useful for evaluating prolonged PTT results. Almost all prolonged PT samples result from factor deficiency, and the mixing test is less useful in this circumstance.

7.4 TT Assay

The thrombin time (TT) assay measures the conversion of fibrinogen to fibrin. It is performed by addition of purified thrombin to patient plasma; the resulting clotting time is a function of fibrinogen concentration and activity. The TT is a screening test for the presence of heparin in a plasma sample. Other causes for prolonged thrombin time include quantitative deficiency of fibrinogen, qualitative abnormality of fibrinogen (dysfibrinogen), elevated levels of fibrin degradation products (FDP), presence of certain paraproteins, and markedly increased levels of fibrinogen (>5 g/l) (Rodgers 2004). If heparin is suspected as the cause of a prolonged TT, heparin presence can be confirmed using heparin assays or by using the reptilase clotting time, which also measures the conversion of fibrinogen to fibrin. The reptilase clotting time is not affected by heparin. Therefore, a plasma sample with a prolonged TT but a normal reptilase clotting time indicates the presence of heparin. For hypofibrinogenemia to prolong a TT, the fibrinogen value will usually be ≤0.7–1 g/l. If the thrombin concentration in the TT assay is more than 3 U/ml, the assay will be less sensitive in detecting abnormalities.

7.5 Fibrinogen Assays

Fibrinogen is a heterodimeric molecule, with each half of the molecule composed of three polypeptide chains: Aα, Bβ, and γ. It is acted upon by thrombin to produce fibrin monomers that polymerize to form fibrin strands, and ultimately a fibrin clot. Circulating fibrinogen molecules are structurally heterogeneous, and not all fibrinogen molecules are capable of participating in clot formation. Therefore, antigenic assays and clot-based assays may return different results depending upon the composition of fibrinogen molecules in a specific patient sample (Lowe et al. 2004). Only clottable fibrinogen is of interest for the purpose of hemostasis screening. Fibrinogen assays available on current automated coagulation analyzers include the Clauss and PT-derived methods. The Clauss method is a modified thrombin time. To initiate clotting, excess thrombin is added to patient plasma that has been diluted with buffer. The clotting time is proportional to the fibrinogen concentration. The dilution with buffer decreases the effects of interfering substances (e.g., heparinoids, FDPs) on the Clauss reaction, relative to the thrombin time assay. However, most manufacturers only claim to inhibit interference from heparin up to a concentration of 1–2 U/ml. Higher concentrations may result in falsely low fibrinogen measurements. FDPs generally result in decreased fibrinogen estimates by the Clauss method, as well, though the degree of interference varies by manufacturer. Some manufacturers include FDP inhibitors in their reagents (Mackie et al. 2003; Lowe et al. 2004). Since Clauss assays have excess thrombin in the reagents, carryover of thrombin to subsequent tests is a potential problem that should be addressed during instrument validation. In the PT-derived assays, the fibrinogen concentration is proportional to the total change in optical signal observed during the PT assay. PT-derived assays have the advantage that a fibrinogen level can be derived directly from the PT assay without additional time or expense. In addition, FDPs generated by thrombolysis do not interfere with these assays (Mackie et al. 2003; Lowe et al. 2004). However, these assays are not recommended for routine use in the clinical laboratory since they have lower precision than the Clauss assays, and suffer from decreased accuracy at low or high fibrinogen concentrations, or when turbid plasma is tested (Mackie et al. 2003). In most circumstances, the PT, PTT, and platelet assays are sufficient for screening bleeding patients. However, in cases where fibrinogen levels may drop precipitously [e.g., disseminated intravascular coagulation (DIC) in obstetric patients], a fibrinogen assay is an essential screening test, since the PT and PTT are relatively insensitive to low levels of fibrinogen.

7.6 D-dimer Assays

D-dimer is formed through the proteolytic action of plasmin on polymerized fibrin that has been cross-linked by factor XIII (Fig. 14). The presence of D-dimer in the circulation is evidence that clot has formed and has been cleaved by plasmin. As such, D-dimer is an effective screening assay for two conditions: DIC and venous thromboembolism (VTE). In DIC, excessive thrombin is formed, resulting in clot formation and activation of the fibrinolytic system that, in turn, produces D-dimer. In VTE, clot forms in the deep venous system of the pelvis and/or proximal, lower extremities secondary to predisposing conditions (Virchow's triad). Clots may then embolize to the lungs. Plasmin cleaves cross-linked fibrin polymers in the clot, releasing D-dimer into the circulation. D-dimer assays are immunoassays with different antibody specificities for the heterogeneous D-dimer fragments produced by the action of plasmin on cross-linked clot (Fig. 14). Therefore, D-dimer levels produced by different assays are generally not interchangeable. The sensitivity of D-dimer assays is judged relative to reference ELISA assays performed in microtiter wells. The manual agglutination assays are the least sensitive tests, while the automated ELISA and immunoturbidimetric assays are the most sensitive, rapid assays. Any of the available assays is suitable for screening for DIC. Many patients, particularly hospitalized patients, have low levels of plasma D-dimer that exceed the reference interval, but they do not have clinical evidence of DIC. Therefore, assay-specific cutoffs should be established to maximize sensitivity and specificity for diagnosis of DIC (Lehman et al. 2004). Sensitive D-dimer assays have been demonstrated to have excellent negative predictive value for diagnosis of VTE, when combined with a pretest probability assessment (Stein et al. 2004). Manual agglutination assays are insufficiently sensitive to be used to rule out VTE.

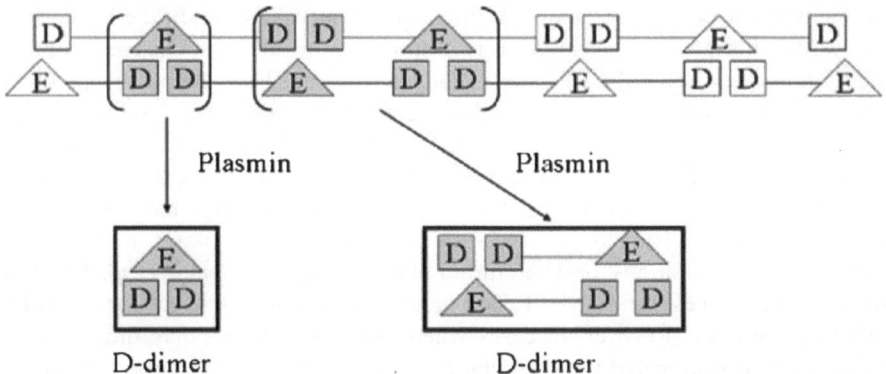

Fig. 14 Proteolytic action of plasmin on cross-linked fibrin polymer producing heterogenous D-dimer molecules

7.7 BT Assay

The bleeding time (BT) test was developed to provide information on platelet function and the likelihood of bleeding with surgery or invasive procedures. However, an extensive literature on the BT has evolved indicating that the test has minimal clinical utility (Rodgers 2004). A clinical outcomes study published in 2001 found that, when the BT test was discontinued at a tertiary medical center, there were no negative impacts, including no increase in procedural or surgical bleeding (Lehman et al. 2001). Figure 15 summarizes an algorithm used to evaluate patients for platelet dysfunction.

8 Conclusions and Future Research Directions

Natural medicines such as enzymes provide a safe, nontoxic therapy for addressing several conditions. One is reminded of the incredible bounty that nature provides when we are able to use a secretion produced by a microorganism found dwelling in the innards of another creature.

Enzymes are not only useful in promoting chemical reactions within the body to ensure its smooth function. These versatile chemicals can also be used to help prevent and treat a number of conditions. It is important, however, that patients be told not to take enzymes on their own, as they can have side-effects involving blood coagulation as well as having negative effects on patients who are pregnant or breastfeeding. Used with proper laboratory testing prior to treatment and with close clinical supervision, however, enzymes can be helpful for patients with clotting disorders, heart conditions, and various types of inflammation.

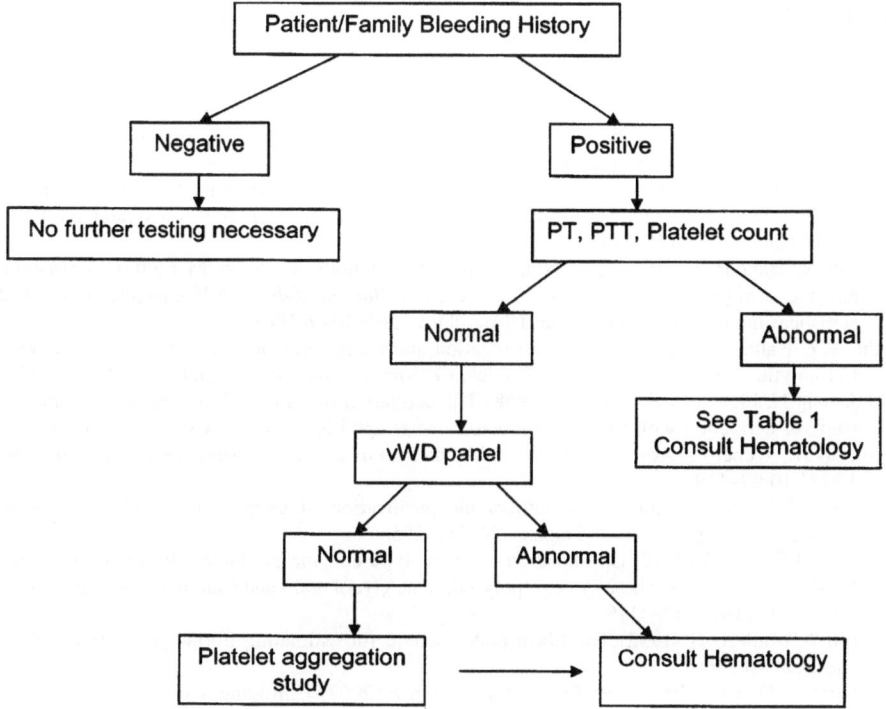

Fig. 15 An algorithm to evaluate patients for possible platelet dysfunction. Patient and family history are important in deciding whether to pursue laboratory evaluation. If history is positive, screening tests of hemostasis (PT, PTT, and platelet count) are used for further evaluation. The vWD panel consists of factor VIII activity, von Willebrand factor antigen, and ristocetin cofactor activity

The microbial fibrinolytic enzymes, especially those from food-grade micro-organisms, have the potential to be developed as functional food additives and drugs to prevent or cure cardiovascular diseases. NK has already been developed as drugs in the market, including Nattokinase NSKSD, Jarrow NattoMax JR-154, and Natto-K. Development of other microbial fibrinolytic enzymes is still ongoing, and much work needs to be done intensively and extensively, especially concerning thrombolytic effects in vivo.

The new trend for improving thrombolytic agents is to increase the efficacy and fibrin specificity, focusing on developing effective targeted thrombolytic agents. Several reports have illustrated successful construction of chimeric proteins in which a thrombus-specific polypeptide or antibody is attached to a plasminogen activator to enhance thrombolytic specificity (Ruppert et al. 2003; Tait et al. 1995). This antibody should be a bifunctional molecule that contains both a highly specific antibody-containing site for concentrating the molecule at the desired target, the thrombus, and an effector site for initiating thrombolysis. These advances indicate the route of future research on microbial fibrinolytic enzymes.

References

Abdel-Fattah AF, Ismail AS (1984) Purification and some properties of pure *Cochliobolus lunatus* fibrinolytic enzyme. Biotechnol Bioeng 26(5):407–411

Abloundi FB, Hagan JJ (1957) Comparison of certain properties of human plasminogen and proactivator. Proc Soc Exp Biol 95:195–200

Adams DS, Griffin LA, Nachajko WR, Reddy VB, Wei CM (1991) A synthetic DNA encoding a modified human urokinase resistant to inhibition by serum plasminogen activator inhibitor. J Biol Chem 266:8476–8482

Agrebi R, Haddar A, Hajji M, Frikha F, Manni L, Jellouli K, Nasri M (2009) Fibrinolytic enzymes from a newly isolated marine bacterium *Bacillus subtilis* A26 characterization and statistical media optimization. Can J Microbiol 55(9):1049–1061

Ahn MY, Hahn BS, Ryu KS (2003) Purification and characterization of a serine protease with fibrinolytic activity from the dung beetles, *Catharsius molossus*. Thromb Res 112:339–347

Alkjaersig N, Fletcher AP, Sherry S (1958) The activation of human plasminogen. II. A kinetic study of activation with trypsin, urokinase, and streptokinase. J Biol Chem 233:86–90

Ambrus JL, Weber FJ, Ambrus CM (1979) Mechanism of action of fibrinolytic enzymes in vivo. J Med 10:99–119

Andreas EM (1990) Affinity chromatographic purification of streptokinase with monoclonal antibodies. Allerg Immunol (Leipzig) 36:277–285

Ashipala OK, He Q (2008) Optimization of fibrinolytic enzyme production by *Bacillus subtilis* DC-2 in aqueous two-phase system (poly-ethylene glycol 4000 and sodium sulfate). Bioresour Technol 99(10):4112–4119

Astrup T, Mullertz S (1952) The fibrin plate method for estimating fibrinolytic activity. Arch Biochem 40:346

Azuaga AI, Dobson CM, Mateo PL, Conejero-Lara F (2002) Unfolding and aggregation during the thermal denaturation of streptokinase. Eur J Biochem 269:4121–4133

Baewald G, Mayer G, Heikel R, Volzke KD, Roehlig R, Decker KL et al (1975) Fermentative production of *Streptococcus* metabolites, especially streptokinase. German Patent DD 111096

Barett AJ (1995) Proteolytic enzymes: aspartic and metallopeptidases. Methods Enzymol 248:183

Barlow GH, Marder VJ (1980) Plasma urokinase levels measured by chromogenic assay after infusions of tissue culture or urinary source material. Throm Res 18:431–437

Barta G (1966) Dyed fibrin plate assay of fibrinolysis. Can J Physiol Pharmacol 44(2):233–240

Batomunkueva BP, Egorov NS (2001) Isolation, purification and resolution of the extracellular proteinase complex of *Aspergillus ochraceus* 513 with fibrinolytic and anticoagulant activities. Microbiology 70(5):519–522

Bayoudh A, Gharsallah N, Chamkha M, Dhouib A, Ammar S, Nasri M (2000) Purification and characterization of an alkaline protease from *Pseudomonas aeruginosa* MNI. J Ind Microbiol Biotechnol 24:291–295

Beldarrain A, Lopez-Lacomba JL, Kutyshenko VP, Serrano R, Cortijo M (2001) Multidomain structure of a recombinant streptokinase: a differential scanning calorimetry study. J Protein Chem 20:9–17

Bell PH, Dziobkowski CT, Englert ME (1974) A sensitive fluorometric assay for plasminogen, plasmin and streptokinase. Anal Biochem 61:200–208

Bernheimer AW, Gillman W, Hottle GA, Pappenheimer AM (1942) An improved medium for the cultivation of hemolytic streptococcus. J Bacteriol 43:495–498

Bicford AF, Sokolow M (1961) Fibinolysis as related to the urea solubiity of fibri. Thronbos Diathes haemorrh (Stuttg) 5:480

Billroth T (1874) Coccobacteria septica, Georg Reimer, Berlin, p 240

Birkedal-Hansen H (1995) Proteolytic remodeling of extracellular matrix. Curr Opin Cell Biol 7:728–735

Blann AD, Landray MJ, Lip GY (2002) An overview of antithrombotic therapy. Br Med J 325:762–765

Blix S (1962) The effectiveness of activators in clot lysis, with special reference to fibrinolytic therapy: a new method for determination of preformed clot lysis. Acta Med Scand 172:386

Bode C, Runge M, Smalling RW (1996) The future of thrombolysis in the treatment of acute myocardial infarction. Eur Heart J 17:55–60

Boxrud PD, Bock PE (2000) Streptokinase binds preferentially to the extended conformation of plasminogen through lysine binding site and catalytic domain interactions. Biochemistry 39:13974–13981

Boxrud PD, Verhamme IMA, Fay WP, Bock PE (2001) Streptokinase triggers conformational activation of plasminogen through specific interactions of the amino-terminal sequence and stabilizes the active zymogen conformation. J Biol Chem 276:26084–26089

Brockway WJ, Castellino FJ (1974) A characterization of native streptokinase and altered streptokinase isolated from a human plasminogen activator complex. Biochemistry 13:2063–2070

Buckell M (1958) The effect of citrate on euglobulin methods of estimating fibrinolytic activity. J Clin Pathol 11:403

Caramelli P, Mutarelli EG, Caramelli B, Tranchesi B, Pileggi F, Scaff M (1992) Neurological complications after thrombolytic treatment for acute myocardial infarction: emphasis on unprecedented manifestations. Acta Neurol Scand 85(5):331–333

Cartwright T (1974) The plasminogen activator of vampire bat saliva. Blood 43:317–326

Castellino FJ (1981) Recent advances in the chemistry of the fibrinolytic system. Chem Rev 81:431–446

Castellino FJ, Sodetz JM, Brockway WJ, Siefring GE (1976) Streptokinase. Methods Enzymol 45:244–257

Chakrabarti R, Fearnley GR (1962) The fibrinolytic potential as a simple measure of spontaneous fibrinolysis. J Clin Pathol 15:228

Chang CT, Fan MH, Kuo FC, Sung HY (2000) Potent fibrinolytic enzyme from a mutant of *Bacillus subtilis* IMR-NK1. J Agric Food Chem 48(8):3210–3216

Chiang CJ, Chen HC, Chao Y, Tzen JTC (2005) Efficient system of artificial oil bodies for functional expression and purification of recombinant nattokinase in *Escherichia coli*. J Agric Food Chem 53(12):4799–4804

Chisti Y (1998a) Biosafety. In: Subramanian G (ed) Bioseparation and bioprocessing: a handbook, vol 2. Wiley-VCH, New York, pp 379–415

Chisti Y (1998b) Strategies in downstream processing. In: Subramanian G (ed) Bioseparation and bioprocessing: a handbook, vol 2. Wiley-VCH, New York, pp 3–30

Chitte RR, Dey S (2000) Potent fibrinolytic enzyme from a thermophilic *Streptomyces megasporus* strain SD5. Lett Appl Microbiol 31(6):405–410

Chitte RR, Dey S (2002) Production of a fibrinolytic enzyme by thermophilic *Streptomyces* species. World J Microbiol Biotechnol 18(4):289–294

Choi HS, Sa YS (2001) Fibrinolytic and antithrombotic protease from *Spirodela polyrhiza*. Biosci Biotechnol Biochem 65:781–786

Choi HS, Shin PH (1998) Purification and partial characterization of a fibrinolytic protease in *Pleurotus ostreatus*. Mycologia 90(4):674–679

Choi NS, Chang KT, Jae Maeng P, Kim SH (2004) Cloning, expression, and fibrin(ogen)olytic properties of a subtilisin DJ- 4 gene from *Bacillus* sp. DJ-4. FEMS Microbiol Lett 236(2):325–331

Choi NS, Yoo KH, Hahm JH, Yoon KS, Chang KT, Hyun BH, Maeng PJ, Kim SH (2005) Purification and characterization of a new peptidase, bacillopeptidase DJ-2, having fibrinolytic activity: produced by *Bacillus* sp. DJ-2 from Doen-Jang. J Microbiol Biotechnol 15(1):72–79

Christensen LR (1945) Streptococcal fibrinolysis: a proteolytic reaction due to serum enzyme activated by streptococcal fibrinolysin. J Gen Physiol 28:363–383

Christensen LR (1949) Methods for measuring the activity of components of the streptococcal fibrinolytic system and Streptococcal desoxyribonuclease. J Clin Invest 28:163–172

Coffey JA, Jennings KR, Dalton H (2001) New antigenic regions of streptokinase are identified by affinity-directed mass spectrometry. Eur J Biochem 268:5215–5221

Collen D (1998a) Engineered staphylokinase variants with reduced immunogenicity. Fibrinol Proteol 12(Suppl 2):59–65

Collen D (1998b) Staphylokinase: a potent, uniquely fibrin-selective thrombolytic agent. Nat Med 4:279–284

Collen D, Van de Werf F (1993) Coronary thrombolysis with recombinant staphylokinase in patients with evolving myocardial infarction. Circulation 87:1850–1853

Collen D, Lijnen HR (1994) Staphylokinase, a fibrin-specific plasminogen activator with therapeutic potential? Blood 84(3):680–686

Collen D, Lijnen HR (2004) Tissue-type plasminogen activator: a historical perspective and personal account. J Thromb Haemost 2(4):541–546

Collen D, DeCock F, Vanlinthout I, Declerck PJ, Lijnen HR, Stassen JM (1992) Comparative thrombolytic and immunogenic properties of staphylokinase and streptokinase. Fibrinolysis 6:232–242

Conejero-Lara K, Parrado J, Azuaga AI, Smith RAG, Ponting CP, Dobson CM (1996) Thermal stability of the three domains of streptokinase studied by circular dichroism and nuclear magnetic resonance. Protein Sci 5:2583–2591

Cui L, Dong MS, Chen XH, Jiang M, Lv X, Yan G (2008) A novel fibrinolytic enzyme from *Cordyceps militaris*, a Chinese traditional medicinal mushroom. World J Microbiol Biotechnol 24:483–489

Davie EW, Neurath H (1953) Identification of the peptide split from trypsinogen during autocatalytic activation. Biochim Biophys Acta 11:442

Davis HC, Karush F, Rudd JH (1965) Effect of amino acids on steady-state growth of a group A hemolytic streptococcus. J Bacteriol 89:421–427

De Renzo EC, Siiteri PK, Hutchings BL, Bell PH (1967) Preparation and certain properties of highly purified streptokinase. J Biol Chem 242:533–542

Deepak V, Kalishwaralal K, Ramkumarpandian S, Babu V, Senthilkumar SR, Sangiliyandi G (2008) Optimization of media composition for Nattokinase production by *Bacillus subtilis* using response surface methodology. Biores Technol 99:8170–8174

Dixon M, Webb EC (1964) Enzymes, 2nd edn. Academic, New York

Duffy MJ (2002) Urokinase plasminogen activator and its inhibitor, PAI-1, as prognostic markers in breast cancer: from pilot to level 1 evidence studies. Clin Chem 48(8):1194–1197

Egorov NS, Kochetov GA, Khaidarova NV (1976) Isolation and properties of the fibrinolytic enzyme from the *Actinomyces thermovulgaris* cultural broth. Mikrobiologiia 45:455–459

Egorov NS, Prianishnikova NI, Al-Nuri MA, Aslanian RR (1985) Streptomyces spheroides M8–2 strain—a producer of extracellular proteolytic enzyme possessing fibrinolytic and thrombolytic action. Naucn Dokl Vyss Sk Biol Nauki 1:77–81

Einarsson M, Skoog B, Forsberg B, Einarsson R (1979) Characterization of highly purified native streptokinase and altered streptokinase after alkaline treatment. Biochim Biophys Acta 568:19–29

Eisenberg PR, Jaffe AS, Stump DC, Collen D, Bovill EG (1990) Validity of enzyme-linked immunosorbent assays of cross-linked fibrin degradation products as a measure of clot lysis. Circulation 82:1159–1168

El-Aassar SA (1995) Production and properties enzyme in solid state cultures of *Fusarium pallidoroseum*. Biotechnol Lett 17(9):943–948

El-Aassar SA, El-Badry HM, Abdel-Fattah AF (1990) The biosynthesis of proteases with fibrinolytic activity in immobilized cultures of *Penicillium chrysogenum* H9. Appl Microbiol Biotechnol 33(1):26–30

Ellis RP, Armstrong CH (1971) Production of capsules, streptokinase, and strepodonase by streptococcus group E. Am J Vet Res 32:349–356

Esch PM, Gerngross H, Fabian A (1989) Reduction of postoperative swelling: objective measurement of swelling of the upper ankle joint in treatment with serrapeptase-a prospective study. Fortschr Med 107:67–68,71–72

Estrada MP, Hernandez L, Perez A, Rodriguez P, Serrano R, Rubier AR et al (1992) High-level expression of streptokinase in *Escherichia coli*. Biotechnology 10:1138–1142

Fearnley GR, Balmforth G, Fearnley E (1957) Evidence of a diurnal fibrinolytic rhythm; with a simple method of measuring natural fibrinolysis. Clin Sci 16:645

Feldman LJ (1974) Streptokinase manufacture (in German). German Patent DE 2354019

Fitzmaurice DA, Blann AD, Lip GY (2002) Bleeding risks of antithrombotic therapy. Br Med J 325:828–831

Flute PT (1960) In: Proceedings of 7th congress European Society of Haematology part II, London, p 894

Fossum S, Hoem NO (1996) Urokinase and non-urokinase fibrinolytic activity in protease-inhibitor-deprived plasma, assayed by a fibrin micro-plate method. Immuno Pharmacol 32:119–121

Fu L, Li RP, Li J, Zhang SM, Zhang YH, Zhao XX, Yang ZX (1997) Study on the fibrinolytic enzyme of *Bacillus subtilis*: selection of liquid fermentation condition. Chin Prog Biotechnol 17(3):31–33

Fujita M, Hong K, Ito Y (1995a) Transport of nattokinase across the rat intestinal tract. Biol Pharm Bull 18(9):1194–1196

Fujita M, Hong K, Ito Y, Fujii R, Kariya K, Nishimuro S (1995b) Thrombolytic effect of nattokinase on a chemically induced thrombosis model in rat. Biol Pharm Bull 18(10):1387–1391

Fujita M, Ito Y, Hong K, Nishimuro S (1995c) Characterization of nattokinase-degraded products from human fibrinogen or crosslinked fibrin. Fibrinolysis 9(3):157–164

Fujita M, Nomura K, Hong K, Ito Y, Asada A, Nishimuro S (1993) Purification and characterization of a strong fibrinolytic enzyme (nattokinase) in the vegetable cheese natto, a popular soybean fermented food in Japan. Biochem Biophys Res Commun 197(3):1340–1347

Galler LI (2000) Streptokinase derivatives with high affinity for activated platelets and methods of their production and use in thrombolytic therapy. US Patent 6087332

Gallimore MJ (1967) Effect of diluents on blood clot lysis. J Clin Pathol 20:234

Gardell SJ, Duong LT, Diehl RE, York JD, Hare TR, Register RB, Jacobs JJ, Dixon RA, Friedman PA (1989) Isolation, characterization, and cDNA cloning of a vampire bat salivary plasminogen activator. J Biol Chem 264:17947–17952

Gase K, Ellinger T, Malke H (1995) Complex transcriptional control of the streptokinase gene of *Streptococcus equisimilis* H46A. Mol Gen Genet 247:749–758

Gidron E, Margalit R, Shalitin Y (1978) A rapid screening test for reduced fibrinolytic activity of plasma: streptokinase activated lysis time. J Clin Pathol 31:54–57

Goldberg AR (1974) Increased protease levels in transformed cells: a casein overlay assay for the detection of plasminogen activator production. Cell 2:95–102

Goldhaber SZ, Bounameaux H (2001) Thrombolytic therapy in pulmonary embolism. Semin Vasc Med 1(2):213–220

Govind NS, Mehta B, Sharma M, Modi VV (1981) Protease and carotenogenesis in *Blakeslea trispora*. Phytochemistry 20:2483–2485

Grafe S, Ellinger T, Malke H (1996) Structural dissection and functional analysis of the complex promoter of the streptokinase gene from *Streptococcus equisimilis* H46A. Med Microbiol Immun 185:11–17

Gupta R, Beg QK, Lorenz P (2002) Bacterial alkaline protease: molecular approaches and industrial application. Appl Microbial Biotechnol 59:15–32

Hartley BS (1960) Proteolytic enzymes. Annu Rev Biochem 29:45–72

Haverkate F, Bradman P (1975) Progress in chemical fibrinolysis and thrombolysis, vol 1. Raven, New York, p 151

Hawkey CM, Stafford JL (1964) A standard clot method for the assay of plasminogen activators, anti-activators, and plasmin. J Clin Pathol 17:175

Health Canada (2000) Cardiovascular disease deaths in Canada. www.hcsc.gc.ca/hpb/lcdc/bcrdd/cardio/cvcpim_e.html.

Healy V, O'Connell J, McCarthy TV, Doonan S (1999) The lysinespecifc proteinase from *Armillaria mellea* is a member of a novel class of metalloendopeptidase located in basidiomycetes. Biochem Biophys Res Commun 262:60–63

Hernandez-Pinzon I, Millan F, Bautista J (1997) Streptokinase recovery by cross-flow microfiltration: study of enzyme denaturation. Biosci Biotechnol Biochem 61:1240–1243

Holmstrom B (1965) Streptokinase assay on large agar diffusion plates. Acta Chem Scand 19:1549–1554

Homma K, Wakana N, Suzuki Y, Nukui M, Daimatsu T, Tanaka E, Tanaka K, Koga Y, Nakajima Y, Nakazawa H (2006) Treatment of natto, a fermented soybean preparation, to prevent excessive plasma vitamin K concentrations in patients taking warfarin. J Nutr Sci Vitaminol 52(5):297–301

Hougie C (1990) Partial thromboplastin time and activated partial thromboplastin time tests: one stage prothrombin time. In: Williams WJ, Beutler E, Erslev AJ, Lichtman MA (eds) Hematology, 4th edn. McGraw-Hill, New York, pp 1766–1770

Howell M (1964) A method for assessing clot lysis. J Clin Pathol 17:310

Hrženjak T, Popović M, Božić T, Grdiša M, Kobrehel Đ, Tiška-Rudman LJ (1998) Fibrinonolytic and anticoagulative activities from the earthworm *Eisenia foetida*. Comp Biochem Physiol. 119B:825–832

Huang TT, Malke H, Ferretti JJ (1989) Heterogeneity of the streptokinase gene in group A streptococci. Infect Immun 57:502–506

Hummel BCW, Schor JM, Buck FF, Boggiano E, DeRenzo EC (1965) Quantitative enzymic assay of human plasminogen and plasmin with azocasein as substrate. Anal Biochem 11:532–547

Hwang CM, Kim DI, Kim JE, Huh SH, Min BG, Park JH, Han JS, Lee BB, Kim YI, Ryu ES, Kim JW (2002) In vivo evaluation of lumbrokinase, a fibrinolytic enzyme extracted from *Lumbricus rubellus*, in a prosthetic vascular graft. J Cardiovasc Surg 43:891–894

Hyun HH, Lee YB, Song KH, Jeon JY, Lee HH (1997) Strain improvement for enhanced production of streptokinase and streptodornase in *Streptococcus* sp. J Microbiol Biotechnol 7:101–106

Ikemura H, Inouye M (1988) In vitro processing of pro-subtilisin in *Escherichia coli*. J Biol Chem 263:12959–12963

International Union of Biochemistry, Molecular Biology (1992) Enzyme nomenclature. Academic, Orlando Fla

Ismail A-MS (1981) Biochemical studies on proteolytic enzymes. Ph.D. thesis faculty of science, Al-Azhar University, Naser, Egypt

Ismail A-MS, Emam SAS, El-Shayeb NMA (2004) Production, properties and in vitro application of novel fibrinolytic enzyme by *Bacillus macerans* 3185. Egypt J Biotechnol 17:454–465

Jackson KW, Tang J (1982) Complete amino acid sequence of streptokinase and its homology with serine proteases. Biochemistry 21:6620–6625

Jeon OH, Moon WJ, Kim DS (1995) An anticoagulant fibrinolytic protease from *Lumbricus rubellus*. J Biochem Mol Biol 28:138–142

Jeong YK, Park JU, Baek H, Park SH, Kong IS, Kim DW, Joo WH (2001) Purification and biochemical characterization of a fibrinolytic enzyme from *Bacillus subtilis* BK-17. World J Microbiol Biotechnol 17:89–92

Jeong YK, Kim JH, Gal SW, Kim JE, Park SS, Chung KT, Kim YH, Kim BW, Joo WH (2004) Molecular cloning and characterization of the gene encoding a fibrinolytic enzyme from *Bacillus subtilis* Strain A1. World J Microbiol Biotechnol 20:711–717

Jespersen J, Astrup T (1983) A study of the fibrin plate assay of fibrinolytic agents. Optimal conditions, reproducibility and precision. Haemost 13:301–315

Jespers L, Vanwetswinkel S, Lijnen HR, Van Herzeele N, Collen D, De Maeyer M (1998) Interface scanning and 3D model of the staphylokinase: plasmin activator complex. Fibrinol Proteol 12:4

Jespers L, Vanwetswinkel S, Lijnen HR, Van Herzeele N, Van Hoef B, Demarsin E, Collen D, De Maeyer L (1999) Structural and functional basis of plasminogen activation by staphylokinase. Thromb Haemost 81:479–485

Johnsen LB, Poulsen K, Kilian M, Petersen TE (1999) Purification and cloning of a streptokinase from *Streptococcus uberis*. Infect Immun 67(3):1072–1078

Johnsen LB, Rasmussen LK, Petersen TE, Etzerodt M, Fedosov SN (2000) Kinetic and structural characterization of a two-domain streptokinase: dissection of domain functionality. Biochemistry 39:6440–6448

Kalisz HK (1988) Microbial proteinases. Adv Biochem Eng Biotechnol 36:1–65

Kaneki M, Hedges SJ, Hosoi T, Fujiwara S, Lyons A, Crean SJ, Ishida N, Nakagawa M, Takechi M, Sano Y, Mizuni Y, Hoshino S, Miyao M, Inoue S, Horiki K, Shiraki M, Ouchi Y, Orino H (2001) Japanese feremented soybean food as the major determinant of the large geographic difference in circulating levels of vitamin K2: possible implications for hip-fracture risk. Nutrition 17:315–321

Kase CS, Pessin MS, Zivin JA, Del Zoppo GJ, Furlan AJ, Buckley JW, Littlejohn JK (1992) Intracranial hemorrhage after coronary thrombolysis with tissue plasminogen activator. Am J Med 92(4):384–390

Kazmi KA, Iqbal MP, Rahbar A, Mehboobali N (2002) Anti-streptokinase titers and response to streptokinase treatment in Pakistani patients. Int J Cardiol 82(3):247–251

Kessner A, Troll W (1976) Fluorometric microassay of plasminogen activators. Arch Biochem Biophys 176:411–416

Kho CW, Park SG, Cho S, Lee DH, Myung PK, Park BC (2005) Confirmation of Vpr as a fibrinolytic enzyme present in extracellular proteins of *Bacillus subtilis*. Protein Expr Purif 39:1–7

Kim JH, Kim YS (1998) Purification and characterization of fibrinolytic enzyme from *Armillaria mellea*. Kor J Mycol 26:583–588

Kim JH, Kim YS (1999) A fibrinolytic metalloprotease from the fruiting bodies of an edible mushroom, *Armillariella mellea*. Biosci Biotechnol Biochem 63:2130–2136

Kim SH, Choi NS (2000) Purification and characterization of subtilisin DJ-4 secreted by *Bacillus* sp strain DJ-4 screened from Doen-Jang. Biosci Biotechnol Biochem 64:1722–1725

Kim W, Choi K, Kim Y (1996a) Purification and characterization of a fibrinolytic enzyme produced from *Bacillus* sp. strain CK 11–4 screened from Chungkook-Jang. Appl Environ Microbiol 62:2482–2488

Kim W, Choi K, Kim Y, Park H, Choi J, Lee Y, Oh H, Kwon I, Lee S (1996b) Purification and characterization of a fibrinolytic enzyme produced from *Bacillus* sp. strain CK 11–4 screened from Chungkook-Jang. Appl Environ Microbiol 62(7):1488–2482

Kim HK, Kim GT, Kim DK, Choi WA, Park SH, Jeong YK, Kong IS (1997) Purification and characterization of a novel fibrinolytic enzyme from *Bacillus* sp. KA38 originated from fermented fish. J Ferment Bioeng 84(4):307–312

Kim DM, Lee SJ, Kim IC, Kim ST, Byun SM (2000) Asp[41]–His[48] region of streptokinase is important in binding to a substrate plasminogen. Thromb Resour 99:93–98

Klein G, Kullich W (2000) Short-term treatment of painful osteoarthritis of the knee with oral enzymes: a randomized, double-blind study versus diclofenac. Chem Drug Invest 19(1):15–23

Kline DL (1971) Thrombosis and bleeding disorders. Academic, NY, p 358

Kline DI, Fishman JB (1961) Improved procedure for the isolation of human plasminogen. J Biol Chem 236:3232–3234

Ko JH, Park DK, Kim IC, Lee SH, Byun SM (1995) High-level expression and secretion of streptokinase in *Escherichia coli*. Biotechnol Lett 17:1019–1024

Ko JH, Yan JP, Zhu L, Qi YP (2004) Identification of two novel fibrinolytic enzymes from *Bacillus subtilis* QK02. Comp Biochem Physiol C Toxicol Pharmacol 137:65–74

Koide A, Suzuki S, Kobayashi S (1982) Preparation of polyethylene glycol-modified streptokinase with disappearance of binding ability towards antiserum and retention of activity. FEBS Lett 143:73–76

Kulisek ES, Holm SE, Johnston KH (1989) A chromogenic assay for the detection of plasmin generated by plasminogen activator immobilized on nitrocellulose using a para-nitroanilide synthetic peptide substrate. Med Anal Biochem 177:78–84

Kulseth MA, Helgeland LA (1993) Highly sensitive chromogenic microplate assay for quantification of rat and human plasminógen. Anal Biochem 210:314–317

Kumada K, Onga T, Hoshino H (1994) The effect of natto possessing a high fibrinolytic activity in human plasma. Igaku to Seibutsugaku 128(3):117–119

Lai YP, Huang J, Wang LF, Li J, Wu ZR (2004) A new approach to random mutagenesis in vitro. Biotechnol Bioeng 86(6):622–627

Laki K, Lorand L (1948) On the solubility of fibrin clots. Science 108:280

Lancefield RC (1933) A serological differentiation of human and other groups of hemolytic streptococci. J Exp Med 57:571–595

Lassen M (1952) Heat denaturation of plasminogen in the fibrin plate method. Acta physiol.scand 27:371–376

Lassen N (1953) Heat denaturation of plasminogen in the fibrin plate method. Acta Physiol Scand 27:371–376

Lassen M (1958) The estimation of fibrinolytic components by means of the lysis time method. Scand J Clin Lab Invest 10:384–389

Lee SH, Jeong ST, Kim IC, Byun SM (1997a) Identification of the functional importance of valine-19 residue in streptokinase by N-terminal deletion and site directed mutagenesis. Biochem Mol Biol Int 41:199–207

Lee SH, Kim IC, Bae KH, Byun SM (1997b) Enhanced production and secretion of streptokinase into extracellular medium in *Escherichia coli* by removal of 13 N-terminal amino acids. Biotechnol Lett 19:151–154

Lee SK, Bae DH, Kwon TJ, Lee SB, Lee HH, Park JH, Heo S, Johnson MG (2001) Purification and characterization of a fibrinolytic enzyme from *Bacillus* sp. KDO-13 isolated from soybean paste. J Microbiol Biotechnol 11(5):845–852

Lee SY, Kim JS, Kim JE, Sapkota K, Shen MH, Kim S, Chun HS, Yoo JC, Choi HS, Kim MK, Kim SJ (2005) Purification and characterization of fibrinolytic enzyme from cultured mycelia of *Armillaria mellea*. Protein Expr Purif 43(1):10–17

Lee JS, Bai HS, Park SS (2006) Purification and characterization of two novel fibrinolytic proteases from mushroom, *Fomitella raxinea*. J Microbiol Biotechnol 16:264–271

Lee KC, Shin JS, Kim BS, Cho H, Kim SY, Lee EB (2007) Antithrombotic effects by oral administration of novel proteinase fraction from earthworm *Eisenia andrei* on venous thrombosis model in rats. Arch Pharm Res 30(4):475–480

Lehman CM, Wilson LW, Rodgers GM (2004) Analytic validation and clinical evaluation of the STA LIATEST immunoturbidimetric D-Dimer assay for the diagnosis of disseminated intravascular coagulation. Am J Clin Pathol 122:178–184

Lehman CM, Blaylock RC, Alexander DP, Rodgers GM (2001) Discontinuation of the bleeding time test without detectable adverse clinical impact. Clin Chem 47:1204–1211

Leonardi MS, Gazzara D, Fava C, Foca A, Mastroeni P (1983) Enzyme-linked immunosorbent assay (ELISA) for streptokinase antibodies. Diagn Immunol 1:64–67

Liu L, Houng A, Tsai J, Chowdhry S, Sazonova I, Reed GL (1999) The fibronectin motif in the NH_2-terminus of streptokinase plays a critical role in fibrin-independent plasminogen activation. Circulation 100(Suppl S):1

Liu BY, Song HY (2002) Molecular cloning and expression of nattokinase gene in *Bacillus subtilis*. Acta Biochim Biophys Sin (Shanghai) 34(3):338–340

Liu JG, Xing JM, Shen R, Yang CL, Liu HZ (2004) Reverse micelles extraction of nattokinase from fermentation broth. Biotechnol Eng J 21:273–278

Liu JG, Xing JM, Chang TS, Ma ZY, Liu HZ (2005) Optimization of nutritional conditions for nattokinase production by *Bacillus natto* NLSSE using statistical experimental methods. Process Biochem 40:2757–2762

Lopez-Sendon J, de Lopez SE, Bobadilla JF, Rubio R, Bermejo J, Delcan JL (1995) Cardiovascular pharmacology (XIII). The efficacy of different thrombolytic drugs in the treatment of acute myocardial infarct. Rev Esp Cardiol 48:407–439

Lowe GDO, Rumley A, Mackie IJ (2004) Plasma fibrinogen. Ann Clin Biochem 41:430–440

Lu F, Sun L, Lu Z, Bie X, Fang Y, Liu S (2007) Isolation and identification of an endophytic strain EJS-3 producing novel fibrinolytic enzymes. Curr Microbiol 54:435–439

Macfarlane RG, Piling J (1946) Observations on fibrinolysis: plasminogen, plasmin, and antiplasmin content of human blood. Lancet 2:562

Mackie IJ, Kitchen S, Machin SJ et al (2003) The haemostasis and thrombosis task force of the British Committee for standards in haematology. Guidelines on fibrinogen assays. Br J Haematol 121:396–404

Maggioni AP, Franzosi MG, Santoro E, White H, Van de Werf F, Tognoni G (1992) The risk of stroke in patients with acute myocardial infarction after thrombolytic and antithrombotic treatment. Gruppo Italiano per lo Studio della Sopravvivenza nell'Infarto Miocardico II (GISSI-2), and the International Study Group. N Engl J Med 327(1):1–6

Maillard C, Berruyer M, Serre CM, Dechavanne M, Deelmas PD (1992) Protein-S, a vitamin K-dependent protein, is a bone matrix component synthesized and secreted by osteoblasts. Endocrinol 130:1599–1604

Malke H (1993) Polymorphism of the streptokinase gene—implications for the pathogenesis of poststreptococcal lomerulonephritis. Zentralbl Bakteriol 278:246–257

Malke H, Ferretti JJ (1984) Streptokinase: cloning, expression and excretion by *Escherichia coli*. Proc Natl Acad Sci U S A 81:3557–3561

Malke H, Roe B, Ferretti J (1985) Nucleotide sequence of the streptokinase gene from *Streptococcus equisimilis* H46A. Gene 34:357–362

Malke H, Steiner K, Gase K, Frank C (2000) Expression and regulation of the streptokinase gene. Methods 21:111–124

Marder VJ (1993) Recombinant streptokinase—opportunity for an improved agent. Blood Coagul Fibrinolysis 4:1039–1040

Markland FS (1998) Snake venoms and the hemostatic system. Toxicon 36:1749–1800

Marsh NA, Gaffney NJ (1977) The rapid fibrin plate-a method for plasminogen activator assay. Thromb Haemostat 38:545–551

Matsuo O, Okada K, Fukao H, Tomioka Y, Ueshima S, Watanuki M et al (1990) Thrombolytic properties of staphylokinase. Blood 76:925–929

Matsubara K, Sumi H, Hori K, Miyazawa K (1998) Purification and characterization of two fibrinolytic enzymes from a marine green alga, *Codium intricatum*. Comp Biochem Physiol Biochem Mol Biol 119:177–181

Matsubara K, Hori K, Matsuura Y, Miyazawa K (1999) A fibrinolytic enzyme from a marine green alga, *Codium latum*. Phytochemistry 52(6):993–999

Matsubara K, Hori K, Matsuura Y, Miyazawa K (2000) Purification and characterization of a fibrinolytic enzyme and identification of fibrinogen clotting enzyme in a marine green alga, *Codium divaricatum*. Comp Biochem Physiol Biochem Mol Biol 125(1):137–143

Mazzone A, Catalani M, Costanzo M, Drusian A, Mandoli A, Russo S, Guarini E, Vesperini G (1990) Evaluation of *Serratia* peptidase in acute or chronic inflammation of otorhinolaryngology pathology: a multicentre, double-blind, randomized trial versus placebo. J Int Med Res 18:379–388

McCoy HE, Broder CC, Lottenberg R (1991) Streptokinases produced by pathogenic group C streptococci demonstrate species-specific plasminogen activation. J Infect Dis 164:515–521

Mihara H, Sumi H, Akazawa K, Yoneds T, Mizumoto H (1983) Fibrinolytic enzyme extracted from the earthworm. J Thromb Haemost 50:258

Mihara H, Sumi H, Yoneta T, Mizumoto H, Ikedo R, Seiki M, Maruyama M (1991) A novel fibrinolytic enzyme extracted from the earthworm *Lumbricus rubellus*. Jpn J Physiol 41(3):461–472

Millar WT, Smith JF (1983) The comparison of solid phase and fibrin plate methods for the measurement of plasminogen activators. Thromb Res 30:431–439

Milochau A, Lassegues M, Valembois P (1997) Purification, characterization and activities of two hemolytic and antibacterial proteins from coelomic fluid of the annelid *Eisenia fetida andrei*. Biochim Biophys Acta 1337:123–132

Mine Y, Wong A, Jiang B (2005) Fibrinolytic enzymes in Asian traditional fermented foods. Food Res Int 38:243–250

Miyata K, Maejima K, Tomoda K, Isono M (1970) Serratia protease part I: purification and general properties of the enzyme. Agric Biol Chem 34(2):310–318

Muller J, Malke H (1990) Duplication of the streptokinase gene in the chromosome of *Streptococcus equisimilis* H46A. FEMS Microbiol Lett 72:75–78

Mullertz S (1954) Effect of carboxylic and amino acids on fibrinolysis produced by plasmin, plasminogen activator, and proteinases. Proc Soc Exp Biol Med 85:326–329

Mundada LV, Prorok M, DeFord ME, Figuera M, Castellino FJ, Fay WP (2003) Structure–function analysis of the streptokinase amino terminus (residues 1–59). J Biol Chem 278:24421–24427

Muramatsu S (1912) On the preparation of natto. *Journal of the College of Agriculture*. Imp Univ Tokyo 5(1):81–94

Nakahama K, Yoshimura K, Marumoto R, Kikuchi M, Lee IS, Hase T, Matsubara H (1986) Cloning and sequencing of *Serratia* protease gene. Nucleic Acids Res 14:5843–5855

Nakajima N, Mihara H, Sumi H (1993) Characterization of potent fibrinolytic enzymes in earthworm, *Lumbricus rubellus*. Biosci Biotech Biochem 57:1726–1730

Nakajima N, Ishihara K, Sugimoto M, Sumi H, Mikuni K, Hamada H (1996) Chemical modification of earthworm fibrinolytic enzyme with human serum albumin fragment and characterization of the protease as a therapeutic enzyme. Biosci Biotech Biochem 60:293–300

Nakamura S, Hashimoto Y, Mikami M, Yamanaka E, Soma T, Hino M, Azuma A, Kudoh S (2003) Effect of the proteolytic enzyme serrapeptase in patients with chronic airway disease. Respirol 8:316–320

Nakamura T, Yamagata Y, Ichishima E (1992) Nucleotide sequence of the subtilisin NAT gene, *aprN*, of *Bacillus subtilis* (natto). Biosci Biotechnol Biochem 56(11):1869–1871

Nakamura T, Yamagata Y, Ichishima E (1992) Nucleotide sequence of the subtilisin NAT gene, *aprN*, of *Bacillus subtilis* (natto). Biosci Biotechnol Biochem 56(11):1869–1871

Narciandi RE, Morbe FJ, Riesenberg D (1996) Maximizing the expression of recombinant kringle 1 (Streptokinase) synthesized in *Escherichia coli*: influence of culture and induction conditions. Biotechnol Lett 18:1261–1266

Nemirovich-Danchenko MM, Alekseeva VN, Lebedeva VV, Shashkova NM, Feigel'man BI, Burovaya FI, Smirnova EM (1985) Streptokinase (in Russian). USSR Patent SU 1147749

Nicolini FA, Nichols WW, Mehta JL, Saldeen TGP, Schofield R, Ross M et al (1992) Sustained reflow in dogs with coronary thrombosis with K2P, a novel mutant of tissue plasminogen activator. J Am Coll Cardiol 20:228–235

Nieuwenhuizen W, Wijngaards G, Groeneverd E (1978) Fluorogenic substrates for sensitive and differential estimation of urokinase and tissue plasminogen activator. Haemost 7:146–149

Nihalani D, Sahni G (1995) Streptokinase contains two independent plasminogen-binding sites. Biochem Biophys Res Commun 217:1245–1254

Nihalani D, Kumar R, Rajagopal K, Sahni G (1998) Role of the amino-terminal region of streptokinase in the generation of a fully functional plasminogen activator complex probed with synthetic peptides. Protein Sci 7:637–648

Nikai T, Mori N, Kishida M, Sugihara H, Tu A (1984) Isolation and biochemical characterization of hemorrhagic toxin F from the venom of *Crotalus atrox*. Arch Biochem Biophys 231:309–319

Ninobe M, Hitomi Y, Fujii S (1980) A sensitive colorimetric assay for various proteases using naphthyl ester derivatives as substrates. J Biochem 87:779–783

Noh KA, Kim DH, Choi NS, Kim SH (1999) Isolation of fibrinolytic enzyme producing strains from kimchi. Kor J Food Sci Technol 31:219–223

Nonaka T, Dohmae N, Hashimoto Y, Takio K (1997) Amino acid sequences of metalloendo-peptidases specific for acyl-lysine bonds from *Grifola frondosa* and *Pleurotus ostreatus* fruiting bodies. J Biol Chem 272:30032–30039

Oden A, Fahlen M (2002) Oral anticoagulation and risk of death: a medical record lineage study. Br Med J 325:1073–1075

Ojalvo AG, Pozo L, Labarta V, Torrens I (1999) Prevalence of circulating antibodies against a streptokinase *C*-terminal peptide in normal blood donors. Biochem Biophys Res Commun 263:454–459

Okada K, Ueshima S, Fukao H, Matsuo O (2001) Analysis of complex formation between plasmin(ogen) and staphylokinase or streptokinase. Arch Biochem Biophys 393:339–341

Ozegowski JH, Gerlach D, Kohler W (1983) Influence of physical parameters on the production of streptococcal extracellular proteins in cultures with stabilized pH: 2: temperature dependence of extracellular protein production. Zentralbl Bakteriol Microbiol Hyg 254:361–369

Paik HD, Lee SK, Heo S, Kim SY, Lee H, Kwon TJ (2004) Purification and characterization of the fibrinolytic enzyme produced by *Bacillus subtilis* KCK-7 from Chungkookjang. J Microbiol Biotechnol 14(4):829–835

Pais E, Alexy T, Holsworth RE Jr, Meiselman HJ (2006) Effects of nattokinase, a pro-fibrinolytic enzyme, on red blood cell aggregation and whole blood viscosity. Clin Hemorheol Microcirc 35(1–2):139–142

Pautov VD, Anufrieva EV, Ananeva TD, Saveleva NV, Taratina TM, Krakovyak MG (1990) Structural dynamic and functional properties of native and modified streptokinase. Mol Biol 24:35–41

Perez N, Urrutia E, Camino J, Orta DR, Torres Y, Martinez Y et al (1998) Hydrophobic interaction chromatography applied to purification of recombinant streptokinase. Biotechnology 10:174–177

Permin PM (1947) Properties of the fibrinokinase–fibrinolysin system. Nature 160:571–572

Peng Y, Zhang YZ (2002a) Isolation and characterization of fibrinolytic enzyme-producing strain DC-4 from Chinese douchi and primary analysis of the enzyme property. Chin High Technol Lett 12:30–34

Peng Y, Zhang YZ (2002b) Cloning and expression in *E. coli* of coding sequence of the fibrinolytic enzyme mature peptide from *Bacillus amyloliquefaciens* DC-4. Chin J Appl Environ Biol 8:285–289

Peng Y, Zhang YZ (2002c) Optimization of fermentation conditions of douchi fibrinolytic enzyme produced by *Bacillus amyloliquefaciens* DC-4. Chin Food Ferment Ind 28:19–23

Peng Y, Huang Q, Zhang RH, Zhang YZ (2003) Purification and characterization of a fibrinolytic enzyme produced by *Bacillus amyloliquefaciens* DC-4 screened from douchi, a traditional Chinese soybean food. Comp Biochem Physiol Biochem Mol Biol 134:45–52

Peng Y, Yang XJ, Xiao L, Zhang YZ (2004) Cloning and expression of a fibrinolytic enzyme (subtilisin DFE) gene from *Bacillus amyloliquefaciens* DC-4 in *Bacillus subtilis*. Res Microbiol 155(3):167–173

Petkov D, Christova E, Karadjova M (1973) Amidase activity of urokinase. I. Hydrolysis of alpha-N-acetyl-L-lysine *p*-nitroanilide. Thromb Diath Haemorrh 29:276–285

Pratap J, Kaur J, Rajamohan G, Singh D, Dikshit KL (1996) Role of *N*-terminal domain of streptokinase in protein transport. Biochem Biophys Res Commun 227:303–310

Pratap J, Rajamohan G, Dikshit KL (2000) Characteristics of glycosylated streptokinase secreted from *Pichia pastoris*: enhanced resistance of SK to proteolysis by glycosylation. Appl Microbiol Biotechnol 53:469–475

Ratnoff OD (1952) Studies on a proteolytic enzyme in human plasma. VIII. The effect of calcium and strontium ions on the activation of the plasma proteolytic enzyme. J Exp Med 96:319

Reddy K, Nagendra N, Markus G (1974) Esterase activities in the zymogen moiety of the streptokinase–plasminogen complex. J Biol Chem 249:4851–4857

Reed GL, Kussie P, Parhamiseren B (1993) A functional analysis of the antigenicity of streptokinase using monoclonal antibody mapping and recombinant streptokinase fragments. J Immunol 150:4407–4415

Reed GL, Liu L, Houng AK, Matsueda LH, Lizbeth H (1998) Mechanisms of fibrin independent and fibrin dependent plasminogen activation by streptokinase. Circulation 98:199

Regnault V, Helft G, Wahl D, Czitrom D, Vuillemenot A, Papouin G et al (2003) Anti streptokinase platelet-activating antibodies are common and heterogeneous. J Thromb Hemost 1:1055–1061

Roberts PS (1958) Measurement of the rate of plasmin action on synthetic substrates. J Biol Chem 232:285–291

Robbins KC (1978) The human plasma fibrinolytic system: regulation and control. Mol Cell Biochem 20(3):149–157

Robinson BR, Liu L, Houng AK, Sazanova IY, Reed GL (2000) The streptokinase beta domain plays a critical role in activator complex formation and substrate docking. Circulation 102:490

Robbins KC, Summaria L (1970) Human plasminogen and plasmin. Methods Enzymol 19:184–199

Robbins KC, Summaria L (1976) Human plasminogen and plasmin. Methods Enzymol 45:257–273

Roche PL, Compeau JD, Schaw ST (1983) A rapid and highly sensitive solid-phase radioassay for plasminogen activators. Thromb Res 31:269–277

Roch P (1979) Protein analysis of earthworm coelomic fluid: I Polymorphic system of the natural hemolysin of *Eisenia fetida* andrei. Devel Comp Immun 3:599–608

Rodgers GM (2004) The diagnostic approach to the bleeding disorders. In: Greer JP, Foerster J, Lukens JN et al (eds) Wintrobe's Clinical Hematology, 11th edn. Williams & Wilkins, Baltimore, pp 1511–1528

Rodriguez P, Hernandez L, Munoz E, Castro A, Fuente JDL, Herrera L (1992) Purification of streptokinase by affinitychromatography on immobilized acylated human plasminogen. BioTechniques 12:424

Rodriguez P, Fuentes D, Munoz E, Rivero D, Orta D, Alburquerque S et al (1994) The streptokinase domain responsible for plasminogen binding. Fibrinolysis 8:276–285

Rosenberger RF, Elsden SR (1960) The yields of *Streptococcus faecalis* grown in continuous culture. J Gen Microbiol 22:726–739

Ruppert C, Markart P, Schmidt R, Grimminger F, Seeger W, Lehr CM, Gunther A (2003) Chemical crosslinking of urokinase to pulmonary surfactant protein B for targeting alveolar fibrin. Thromb Haemost 89:53–64

Saksela O (1981) Radial caseinolysis in agarose: a simple method for detection of plasminogen activators in the presence of inhibitory substances and serum. Anal Biochem 111:276–282

Sazonova IY, Houng AK, Chowdhry SA, Reed GL (2000) Mechanism of action of a novel *Streptococcus uberis* plasminogen activator (SUPA). Circulation 102:489

Seo JH, Lee SP (2004) Production of fibrinolytic enzyme from soybean grits fermented by *Bacillus firmus* NA-1. J Med Food 7(4):442–449

Sharma S, Aneja MK, Mayer J, Schloter M, Munch JC (2004) RNA fingerprinting of microbial community in the rhizosphere soil of grain legumes. FEMS Microbiol Lett 240:181–186

Shemanova F, Postnikova TM, Rostov G (1995) Enzyme immunoassay of antistreptokinase. Klin Lab Diagn 2:26–29 (in Russian)

Sherry S, Troll W (1954) The action of thrombin on synthetic substrates. J Biol Chem 208:95–105

Sherry S, Alkjaersig N, Fletcher AP (1964) Assay of urokinase preparations with the synthetic substrate acetyl-L-lysine methyl ester. J Lab Clin Med 64:145–153

Sherry S, Alkjaersig N, Fletcher AP (1966) Activity of plasmin and streptokinase-activator on substituted arginine and lysine esters. Throm Diath Haemorrh 16:18–31

Sherry S, Lindemeyer RI, Fletcher AP, Alkjaersig N (1959) Studies on enhanced fibrinolytic activity in man. J Clin Invest 38:810

Shi GY, Chang BI, Chen SM, Wu DH, Wu HL (1994) Function of streptokinase fragments in plasminogen activation. Biochem J 304:235–241

Shi GY, Chang BI, Su SW, Young KC, Wu DH, Chang LC et al (1998) Preparation of a novel streptokinase mutant with improved stability. Thromb Hemost 79:992–997

Silverstein M (1975) The determination of human plasminogen using Na-CBZ-L-lysin p-nitrophenyl ester as substrate. Anal Biochem 65:500–506

Silverthorn AC, Ober WC, Garrison CW (1998) Human physiology: an integrated approach. Prentice-Hall Inc, New York, pp 465–471

Smith RAG, Dupe RJ, English PD, Green J (1981) Fibrinolysis with acyl-enzymes-a new approach to thrombolytic therapy. Nature 290:505–508

Somerville DA (1972) A technique for demonstrating fibrinolysis by cutaneous bacteria. J Clin Pathol 25:740–741

Stein PD, Hull RD, Patel KC et al (2004) D-Dimer for the exclusion of acute venous thrombosis and pulmonary embolism. Ann Intern Med 140:589–602

Stuebner K, Boschke E, Wolfe K-H, Langer J (1991) Kinetic analysis and modeling of streptokinase fermentation. Acta Biotechnol 11:467–477

Suh H, Kim KH, Kim SS, Han MH (1984) Culture conditions of *Streptococcus* sp. for streptokinase production. Sanop Misaengmul Hakhoechi 12(3):224–231

Sumi H, Nakajima N, Mihara H (1993) A very stable and potent fibrinolytic enzyme found in earthworm *Lumbricus rubellus* autolysate. Comp Biochem Physiol 106(B):763–766

Sumi H, Nakajima N, Yatagai C (1995) A unique strong fibrinolytic enzyme (datsuwokinase) in skipjack "Shiokara", a Japanese traditional fermented food. Comp Biochem and Physiol 112:543–547

Sumi H, Hamada H, Tsushima H, Mihara H, Muraki H (1987) A novel fibrinolytic enzyme (nattokinase) in the vegetable cheese Natto; a typical and popular soybean food in the Japanese diet. Experientia 43(10):1110–1111

Sumi H, Hamada H, Nakanishi K, Hiratani H (1990) Enhancement of the fibrinolytic activity in plasma by oral administration of nattokinase. Acta Haematol 84(3):139–143

Sumi H, Hamada H, Tsushima H, Mihara H, Muraki H (1987) A novel fibrinolytic enzyme (nattokinase) in the vegetable cheese Natto; a typical and popular soybean food in the Japanese diet. Experientia 43(10):1110–1111

Sumi H, Hamada H, Mihara H, Nakanishi K, Hiratani H (1989) Fibrinolytic effect of the Japanese traditional food natto (nattokinase). Thromb Haemost 62(1):549

Sumi H, Nakajima N, Mihara H (1992) In vitro and in vivo fibrinolytic properties of nattokinase. Thromb Haemost 89:1267

Sun T, Liu BH, Li P, Liu DM, Li ZH (1998) New solid-state fermentation process for repeated batch production of fibrinolytic enzyme by *Fusarium oxysporum*. Process Biochem 33(4):419–422

Sundram V, Nanda JS, Rajagopal K, Dhar J, Chaudhary A, Sahni G (2003) Domain truncation studies reveal that the streptokinase–plasmin activator complex utilizes long range protein–protein interactions with macromolecular substrate to maximize catalytic turnover. J Biol Chem 278:30569–30577

Suttie LW, Machlin LJ (1991) Vitamin K In: Handbook of vitamins, 2nd edn. Marcel Dekker, New York and Basel, pp 145–188

Suzuki Y, Kondo K, Ichise H, Tsukamoto Y, Urano T, Umemura K (2003a) Dietary supplementation with fermented soybeans suppresses intimal thickening. Nutrition 19:261–264

Suzuki Y, Kondo K, Matsumoto Y, Zhao BQ, Otsuguro K, Maeda T, Tsukamoto Y, Urano T, Umemura K (2003b) Dietary supplementation of fermented soybean, natto, suppresses intimal thickening and modulates the lysis of mural thrombi after endothelial injury in rat femoral artery. Life Sci 73:1289–1298

Swenson S, Markland FS Jr (2005) Snake venom fibrin(ogen)olytic enzymes. Toxicon 45:1021–1039

Tachibana M, Mizukoshi O, Harada Y, Kawamoto K, Nakai Y (1984) A multi-centre, double-blind study of serrapeptase versus placebo in postantrotomy buccal swelling. Pharmatherapeutica 3:526–530

Tait JF, Engelhardt S, Smith C, Fujikawa K (1995) Prourokinaseannexin V chimeras. construction, expression, and characterization of recombinant proteins. J Biol Chem 270(37):21594–21599

Tang Y, Liang DC, Jiang T, Zhang JP, Gui LL, Chang WR (2002) Crystal structure of earthworm fibrinolytic enzyme component A: revealing the structural determinants of its dual fibrinolytic activity. J Mol Biol 321:57–68

Tao S, Peng L, Beihui L, Deming L, Zuohu L (1997) Solid state fermentation of rice chaff for fibrinolytic enzyme production by Fusarium oxysporum. Biotechnol Lett 19(5):465–467

Tao S, Peng L, Beihui L, Deming L, Zuohu L (1998) Successive cultivation of Fusarium oxysporum on rice chaff for economic production of fibrinolytic enzyme. Bioprocess Eng 18(5):379–381

Taylor FB, Botts J (1968) Purification and characterization of streptokinase with studies of streptokinase activation of plasminogen. Biochemistry 7:232–242

Tillett WS, Garner RL (1933) Fibrinolytic activity of hemolytic streptococci. J Exp Med 58:485–502

Troll W, Sherry S, Wachman J (1954) The action of plasmin on synthetic substrates. J Biol Chem 208:85–93

Tomar RH (1968) Streptokinase: preparation, comparison with streptococcal proteinase, and behavior as a trypsin substrate. Proc Soc Exp Biol Med 127:239–244

Torrens I, Ojalvo AG, Seralena A, Hayes O, de la Fuente J (1999) A mutant streptokinase lacking the C-terminal 42 amino acids is less immunogenic. Immunol Lett 70:213–218

Troll W, Sherry S, Wachman J (1954) The action of plasmin on synthetic substrates. J Biol Chem 208:85–93

Tough J (2005) Thrombolytic therapy in acute myocardial infarction. Nurs Stand 19(37):55–64

Turpie AG, Chin BŠ, Lip GY (2002) Venous thromboembolism: treatment strategies. Br Med J 325:948–950

Urano T, Ihara H, Umemura K, Suzuki Y, Oike M, Akita S, Tsukamoto Y, Suzuki I, Takada A (2001) The profibrinolytic enzyme subtilisin NAT purified from Bacillus subtilis cleaves and inactivates plasminogen activator inhibitor type 1. J Biol Chem 276:24690–24696

Unkeless J, Gardon S, Reich E (1974) Secretion of plasminogen activator by stimulated macrophages. J Exp Med 139:834–850

Wakeham N, Terzyan S, Zhai PZ, Loy JA, Tang J, Zhang XC (2002) Effects of deletion of streptokinase residues 48–59 on plasminogen activation. Protein Eng 15:753–761

Walker ID, Davidson JF (1985) Blood coagulation and haemnostasis—a practical guide, 3rd edn. Churchill Livingstone, Edinburgh, p 229

Wang X, Lin X, Loy JA, Tang J, Zhang XC (1998) Crystal structure of the catalytic domain of human plasmin complexed with streptokinase. Science 281:1662–1665

Wang SG, Reed GL, Hedstrom L (1999) Deletion of Ile1 changes the mechanism of streptokinase: evidence for the molecular sexuality hypothesis. Biochemistry 38:5232–5240

Wang JD, Narui T, Kurata H, Taeuchi K, Hashimoto T, Okuyama T (1989) Hematological studies on naturally occurring substances II. Eject of animal crude drugs on blood coagulation and fibrinolysis systems. Chem Pharm Bull 37:2236–2238

Wang J, Wang M, Wang Y (1999a) Purification and characterization of a novel fibrinolytic enzyme from Streptomyces spp. Chin J Biotechnol 15(2):83–89

Wang SG, Reed GL, Hedstrom L (1999b) Deletion of Ile1 changes the mechanism of streptokinase: evidence for the molecular sexuality hypothesis. Biochemistry 38:5232–5240

Wang F, Wang C, Li M, Cui L, Zhang J, Chang W (2003) Purification, characterization and crystallization of a group of earthworm fibrinolytic enzymes from Eisenia fetida. Biotechnol Lett 25:1105–1109

Wang C, Ji B, Li B, Ji H (2006a) Enzymatic properties and identification of a fibrinolytic serine protease purified from *Bacillus subtilis* DC33. World J Microbiol Biotechnol 22:1365–1371

Wang SL, Kao TY, Wang CL, Yen YH, Chern MK, Chen YH (2006b) A solvent stable metalloprotease produced by *Bacillus* sp. TKU004 and its application in the deproteinization of squid pen for β-chitin preparation. Enzyme Microb Technol 39:724–731

Wang SH, Cheng Z, Yang YL, Miao D, Bai MF (2008) Screening of a high fibrinolytic enzyme producing strain and characterization of the fibrinolytic enzyme produced from *Bacillus subtilis* LD-8547. World J Microbiol Biotechnol 24:475–482

Wang S, Chen H, Liang T, Lin Y (2009) A novel nattokinase produced by *Pseudomonas* sp. TKU015 using shrimp shells as substrate. Process Biochem 44:70–76

Westlund LE, Andersson LO (1991) Variables influencing the clot lysis assay of streptokinase. Thromb Res 64:713–721

Wheatley DJ (2002) Coronary artery surgery at the dawn of the 21st century. J R Coll Surg Edinb 47(4):608–612

White J (2005) Snake venoms and coagulopathy. Toxicon 45:951–967

Wong SL (1995) Advances in the use of *Bacillus subtilis* for the expression and secretion of heterologous proteins. Curr Opin Biotechnol 6:517–522

World Health Organization (2000) The World Health Report in 2000 C F, Wong AHK (2003) A novel fibrinolytic enzyme from fermented shrimp paste. M.Sc. thesis, University of Guelph

World Health Organization (2001) The World Health Report 2001 C F, Wong AHK (2003) A novel fibrinolytic enzyme from fermented shrimp paste. M.Sc. thesis, University of Guelph

Wong SL, Ye RQ, Nathoo S (1994) Engineering and production of streptokinase in *Bacillus subtilis* expression–secretion system. Appl Environ Microbiol 60:517–523

Wu KK, Thiagarajan P (1996) Role of endothelium in thrombosis and hemostasis. Annu Rev Med 47:315–331

Wu XC, Ye RQ, Duan YJ, Wong S-L (1998) Engineering of plasmin-resistant forms of streptokinase and their production in *Bacillus subtilis*: streptokinase with longer functional half-life. Appl Environ Microbiol 64:824–829

Wu DH, Shi GY, Chuang WJ, Hsu JM, Young KC, Chang CW et al (2001) Coiled coil region of streptokinase gamma domain is essential for plasminogen activation. J Biol Chem 276:15025–15033

Xiao L, Zhang RH, Peng Y, Zhang YZ (2004) Highly efficient gene expression of a fibrinolytic enzyme (subtilisin DFE) in *Bacillus subtilis* mediated by the promoter of α-amylase gene from *Bacillus amyloliquefaciens*. Biotechnol Lett 26:1365–1369

Xiao-Lan L, Lian-Xiang D, Fu-Ping L, Xi-Qun Z, Jing X (2005) Purification and characterization of a novel fibrinolytic enzyme from *Rhizopus chinensis* 12. Appl Microbiol Biotechnol 67(2):209–214

Xu YH, Liang GD, Sun ZJ, Chem F, Fu SH, Chai YB, Hou YD (2002) Cloning and expression of the novel gene-*pV242* of earthworm fibrinolytic enzyme. Prog Biochem Biophys 29:610–614

Yang J, Ru B (1997) Purification and characterization of an SDS activated fibrinolytic enzyme from *Eisenia fetida*. Comp Biochem Physiol 118B(3):623–631

Yoon SJ, Yu MA, Sim GS, Kwon ST, Hwang JK, Shin JK, Yeo IH, Pyun YR (2002) Screening and characterization of microorganisms with fibrinolytic activity from fermented foods. J Microbiol Biotechnol 12(4):649–656

Yazdani SS, Mukherjee KJ (1998) Overexpression of streptokinase using a fed-batch strategy. Biotechnol Lett 20:923–927

Yazdani SS, Mukherjee KJ (2002) Continuous culture studies on the stability and expression of recombinant streptokinase in *Escherichia coli*. Bioprocess Biosyst Eng 24:341–346

Young KC, Shi GY, Chang YF, Chang BI, Chang LC, Lai MD et al (1995) Interaction of streptokinase and plasminogen-studied with truncated streptokinase peptides. J Biol Chem 270:29601–29606

Zhai P, Wakeham N, Loy JA, Zhang XC (2003) Functional roles of streptokinase C-terminal flexible peptide in active site formation and substrate recognition in plasminogen activation. Biochemistry 42:114–120

Zhang XW, Sun T, Huang XN, Liu X, Gu DX, Tang ZQ (1999) Recombinant streptokinase production by fed-batch cultivation of *Escherichia coli*. Enzyme Microb Technol 24:647–650

Zhang RH, Xiao L, Peng Y, Wang HY, Bai F, Zhang YZ (2005) Expression and characteristics of a novel fibrinolytic enzyme (subtilisin DFE) in *Escherichia coli*. Lett Appl Microbiol 41:190–195

Zheng ZL, Zuo ZY, Liu ZG, Tsai KC, Liu AF, Zou GL (2005) Construction of a 3D model of nattokinase, a novel fibrinolytic enzyme from *Bacillus natto*. A novel nucleophilic catalytic mechanism for nattokinase. J Mol Graph Model 23:373–380

Zhu X, Ohta Y, Jordan F, Inouye M (1989) Pro-sequence of subtilisin can guide the refolding of denatured subtilisin in an intermolecular process. Nature 339:483–484

Zimmerman M, Quigley JP, Ashe B, Dron C, Goldfarb R, Troll W (1978) Direct fluorescent assay of urokinase and plasminogen activators of normal and malignant cells: kinetics and inhibitor profiles. Proc Natl Acad Sci U S A 75:750–753